GRAND COMPLICATION

VOL

CHRISTOPHE CLARET

GRAND COMPLICATIONS®

THE ORIGINAL ANNUAL OF THE WORLD'S WATCH COMPLICATIONS AND MANUFACTURERS®

SPECIAL ALTERNATIVE DISPLAY EDITION

First published in the United States in 2014 by

TOURBILLON INTERNATIONAL
A MODERN LUXURY COMPANY

11 West 25th Street, 8th Floor
New York, NY 10010
Tel: +1 (212) 627-7732   Fax +1 (312) 274-8418
www.modernluxury.com/watches

Caroline Childers
PUBLISHER

Michel Jeannot
EDITOR IN CHIEF

Lew Dickey
CHIEF EXECUTIVE OFFICER

Michael Dickey
PRESIDENT

John Dickey
EXECUTIVE VICE PRESIDENT AND CO-CHIEF OPERATING OFFICER

Jon Pinch
EXECUTIVE VICE PRESIDENT AND CO-CHIEF OPERATING OFFICER

JP Hannan
CHIEF FINANCIAL OFFICER

Richard Denning
GENERAL COUNSEL

In association with **RIZZOLI** INTERNATIONAL PUBLICATIONS, INC.

300 Park Avenue South, New York, NY 10010

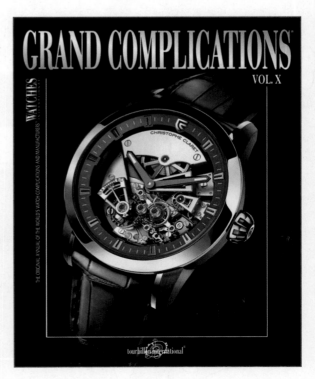

ISBN: 978-0-8478-4302-2

DISCLAIMER: THE INFORMATION CONTAINED IN WATCHES INTERNATIONAL 2014 HAS BEEN PROVIDED BY THIRD PARTIES. WHILE WE BELIEVE THESE SOURCES TO BE RELIABLE, WE ASSUME NO RESPONSIBILITY OR LIABILITY FOR THE ACCURACY OF TECHNICAL DETAILS CONTAINED IN THIS BOOK.

EVERY EFFORT HAS BEEN MADE TO LOCATE THE COPYRIGHT HOLDERS OF MATE-RIALS PRINTED IN THIS BOOK. SHOULD THERE BE ANY ERRORS OR OMISSIONS, WE APOLOGIZE AND SHALL BE PLEASED TO MAKE ACKNOWLEDGMENTS IN FUTURE EDITIONS.

PRINTED IN ITALY

COVER: MAESTOSO (CHRISTOPHE CLARET)

Breguet, the innovator.
Classique Hora Mundi – 5717

An invitation to travel across the continents and oceans illustrated on three versions of the hand-guilloché lacquered dial, the Classique Hora Mundi is the first mechanical watch with an instant-jump time-zone display. Thanks to a patented mechanical memory based on two heart-shaped cams, it instantly indicates the date and the time of day or night in a given city selected using the dedicated pushpiece. History is still being written…

RICHARD MILLE

A RACING MACHINE ON THE WRIST

RM 50-01 CHRONOGRAPH G-SENSOR
LOTUS F1 TEAM-ROMAIN GROSJEAN

Manual winding tourbillon movement
Power reserve: circa 70 hours
Bridges and baseplate in grade 5 titanium
Chronograph (column wheels in grade 5 titanium)
G-force indicator
Fast rotating barrel (6 hours per revolution instead of 7.5 hours)
Modular time setting mechanism fitted against the case back
Extreme finishing for the movement
Balance wheel: Glucydur, with 2 arms and 4 setting screws,
moment of inertia 10 mg·cm², angle of lift 53°
Frequency: 21,600 vph (3 Hz)
Case in NTPT® carbon
Torque limiting crown
Hand-polished beveling
Microblasted milled sections

Limited edition of 30 pieces

Letter from the President

The Ultimate Playground

Like a perfectly assembled domino track, propelled into continuous motion by the push of a single piece, the complicated timepiece captures man's attention as if the result were an enigma solved only by its flawless execution. Our fascination with the moving object is nothing new. It is neither the question, nor the answer that captivates the most visceral parts of our minds.

Surely, as the perpetual calendar nears its complete synchronized metamorphosis at the turn of a new year, it is not the mystery of its outcome that enthralls its attentive observer.

This newest volume of *Grand Complications* presents the most sophisticated displays of modern micro-mechanical prowess. It showcases the world's finest triumphs of automated masterpieces, perfectly tuned machines that perform their theater on the wrist.

Astronomical complications replicate the motions of the universe in a space smaller than the opening of a telescope. Tourbillons connect us to Sir Isaac Newton's glorious discoveries on universal gravitation. Equations of time remind us that each day varies in actual length from the last, that Earth's relationship with the sun holds many asymmetrical nuances.

Just as antique automatons were admired for the genius of their self-operating movements and not for their industrial functions, the timekeeping complications displayed in this publication hold their fascination not in the what, but in the how. These mechanical playgrounds permit us to witness the ingenuity of mankind. As each gear and each component plays off the next to achieve a stunning choreography, time is neither the question, nor the answer: it is an exhibition of a superb intellectual pursuit, a rejection of the impossible.

Michael Dickey

We have changed 16 out of 405 parts and added another 46. For the benefit of an extra power reserve.

GLASHÜTTE I/SA

The power-reserve indicator of the DATOGRAPH UP/DOWN reveals that its autonomy has been extended to 60 hours. Moreover, it is endowed with a proprietary freely oscillating hairspring and a balance wheel with eccentric poising weights for superior rate accuracy. And while no fewer than 62 parts were reworked, Lange's watchmakers preserved the proven design features. They invested a considerable amount of work that reveals itself only to aficionados - an A. Lange & Söhne custom that has always been appreciated. **www.lange-soehne.com**

A. Lange & Söhne · 252 Worth Avenue · Palm Beach, FL 33480 · Tel. 561 833 0803

cellini

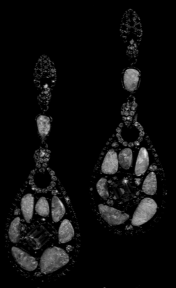

A talent for breaking watch brands in America for nearly 40 years is why Cellini Jewelers is the place to see what's next.

Its extensive collection brings together many of today's top watchmakers, but over the years Cellini has also served as a crucial launching pad for many prominent brands. Before their popularity soared among collectors, firms like A. Lange & Söhne, Breitling, Franck Muller and Hublot all found a home at Cellini. Even today Cellini remains an influential tastemaker by focusing on watches that are as innovative as they are rare, including timepieces by De Bethune, Greubel Forsey, HYT and Ludovic Ballouard.

Cellini continues to stay ahead of the curve this year by introducing the outstanding work of two gifted watchmakers, Robert Greubel and Stephen Forsey. Their firm, Greubel Forsey, is at the forefront of horological innovation, using meticulous research and development to improve timekeeping precision with original inventions like the Tourbillon 24 Secondes and Quadruple Tourbillon. And while tourbillons are something of an obsession at Greubel Forsey, the firm proves its versatility with the GMT. Technically brilliant with aesthetics to match, the watch is instantly recognizable thanks to the globe rotating on its dial.

Whether you're drawn to the aesthetics or mechanics of watchmaking, Cellini has something to tempt you with one of the most extensive selections of mechanical timepieces from the world's best watchmakers. That incredible depth allows the company to create an unparalleled experience for people who are passionate about watches.

What it does best, according the Cellini founder Leon Adams, is use that range to help someone find the right watch. He explains, "If you like a particular complication or style, we line up different models from various brands so you can weight the options and judge for yourself what feels right. You can't find that anywhere else."

▲ **EMERALD TOURMALINE DROP EARRINGS**
Rare black opals surround tourmalines in these drop earrings.

▲ **EMERALD OPAL RING**
Half-moon opals create a ring of fiery iridescence around the emerald cabochon in this ring.

◄ **GMT**
Greubel Forsey's GMT includes a rotating, three-dimensional globe.

SPARKLING PERSONALITY

But Cellini is known for more than just watches. Its boutiques are also ranked among Manhattan's finest jewelers thanks to impeccably high standards reflected in everything from its unrivaled selection and superior quality to the attentive experts who are ready to guide you through Cellini's sparkling universe.

Its collection combines a wide range of styles with an extraordinary depth that makes anything possible, whether you want subtle or show-stopper. You'll not only find diamonds in every hue, but also impressive strands of Tahitian and South Sea pearls, as well as magnificent emeralds, rubies and sapphires in one-of-a-kind handmade settings.

Just this year, Cellini turned the spotlight on opals with several new pieces that harness the glowing gemstone's colorful pyrotechnics. One head-turning creation is a pair of drop earrings that feature vibrant tourmalines surrounded by black opals. This color opal is prized for its extreme rarity and also its dark color, which is the ideal backdrop for the gemstone's signature light show. There's also a floral-themed ring featuring half-moon shaped opals. They form a fiery border around a cabochon emerald, making this chic design irresistible.

Cellini also expanded its popular Black and White Diamond Collection this year with several distinctive pieces. One of the most impressive is a zipper necklace with an interlocking pattern of black diamonds and clusters of brilliant white diamonds. With 47.2 total carats, it's red-carpet ready. The same is true of the diamond bracelet. The links in its open design are blackened 18-karat gold, giving the elegance a dark edge while amplifying the radiance of the brilliant white diamonds.

Come to Cellini and discover the unrivaled selection of fine jewelry and collectible timepieces.

1815 RATTRAPANTE PERPETUAL CALENDAR
(A. Lange & Söhne)

ROYAL OAK OFFSHORE CHRONOGRAPH
(Audemars Piguet)

ROTONDE DE CARTIER MYSTERIOUS DOUBLE TOURBILLON
(Cartier)

RM 27-01
(Richard Mille)

Hotel Waldorf-Astoria
301 Park Avenue at 50th Street
New York • NY 10022
212-751-9824

509 Madison Avenue at 53rd Street
New York • NY 10022
212-888-0505

800-CELLINI
www.CelliniJewelers.com

AUTHORIZED RETAILER

A. LANGE & SÖHNE
AUDEMARS PIGUET
BELL & ROSS
BVLGARI
CARTIER
CHOPARD
DE BETHUNE
FRANCK MULLER
GIRARD-PERREGAUX
GIULIANO MAZZUOLI
GREUBEL FORSEY
H. MOSER & CIE.
HUBLOT
HYT
IWC
JAEGER-LECOULTRE
JEAN DUNAND
LUDOVIC BALLOUARD
MAÎTRES DU TEMPS
PARMIGIANI FLEURIER
PIAGET
RICHARD MILLE
ROGER DUBUIS
ULYSSE NARDIN
VACHERON CONSTANTIN
ZENITH

◄ DIAMOND ZIPPER NECKLACE
This 18-karat gold zipper-motif necklace impresses with black and white diamonds that total 47.2 carats.

▼ WHITE DIAMOND BRACELET
Blackened 18-karat gold provides contrast for the brilliant white diamonds in this bracelet.

▼ DB28 SKYBRIDGE
The blued-titanium dial of De Bethune's DB28 Skybridge is sprinkled with gold and diamonds.

GPHG

GRAND PRIX D'HORLOGERIE DE GENÈVE

— 2012 —

Best Ladies' Watch Prize

LA MONTRE PREMIÈRE

CHANEL

FLYING TOURBILLON

High feminine complication, this flying tourbillon decorated with the motif of the camellia, a tribute to Mademoiselle Chanel's favorite flower, beats away discreetly and almost secretly at the heart of the Première watch. Having no upper bridge, the carriage decorated with a camellia appears to be rotating in a weightless state. Limited edition of 20 numbered pieces. 18-carat white gold, set with 228 diamonds (~7.7 carats).

Letter from the Publisher

Everything old is new again

Everything old is new again in haute horology, with real innovation thrown in for good measure! The year 2014 promises to be a very eventful one for the industry, with the advent or confirmation of a few pronounced trends. First of all—and this may be just the first few tentative steps—watches seem to be regaining more reasonable dimensions after the excesses of the last few years. Of course, no  one should expect to see tiny timepieces on big, strong wrists, but the trend towards less exaggerated sizes is real, and will become more pronounced in the months to come. Don't misunderstand: the sizes from the 1950s and '60s will not be the standard again, but watchmakers and their clients now prefer watches with slightly smaller dimensions. Everyone, or almost everyone, will fall into step on this one.

The other outstanding theme is the ever-increasing presence of "métiers d'art," those time-honored horological embellishments. After extraordinary mechanical developments over the last decade, incorporated into hyper-complicated models, watchmakers are developing a new field of expression with these artistic flourishes. January's SIHH and BaselWorld were occasions for showing off the return of these professions, which were inseparably linked with haute horology a century or two ago. The renaissance is in full swing, and is so strong a trend that even brands with no experience in the field make full-throated declarations to the public that their mission is inextricably tied to their wish to perpetuate these nearly disappeared artistic traditions. As amusing as these protestations may be, one thing is certain: the more watchmakers emphasize métiers d'art, the more assured the tradition's continuation. However, the enthusiasm for métiers d'art goes beyond historical recognition; in fact, in addition to reviving age-old artistry, certain brands are now inventing their own, modern métiers d'art. Contemporary watchmaking is perpetually reinventing itself on every level!

Caroline Childers

Villeret Collection

HUBLOT

THE ART OF FUSION

Proceeds obtained from the
licensing of the Ayrton Senna's
image is 100% invested
in the educational projects
developed by the Ayrton Senna
Institute throughout Brazil.

MP06-Senna.
A new shape. A powerful barrel-shaped design
in keeping with the Hublot stylistic codes. Equipped
with a skeleton tourbillon movement with a 5-day power
reserve. Entirely manufactured by Hublot. A limited
edition of only 41 numbered pieces in tribute
to the 41 victories of the legendary Brazilian
icon Ayrton Senna.

HUBLOT

THE ART OF FUSION

OFFICIAL WATCH
SCUDERIA FERRARI

MP05-LaFerrari.
A truly exceptional watch.
A world record-holder.
50-day power reserve and a high-tech
design developed with Ferrari.
Limited edition of 50 pieces.

Letter from the Editor in Chief

Ultra-thin—a complication in its own right

When we speak of horological complications, the first functions that come to mind are invariably the tourbillon, the minute repeater or the perpetual calendar, maybe even the chronograph—that is, a list of mechanisms developed in the early days of horology that today's watchmakers continually "reinvent" and interpret as they please. These very pages demonstrate the tenacity of that concept. There is, however, an approach that one might consider a pure style exercise, but which nonetheless represents a complication all its own: the extra-thin movement. The quest for the thinnest, and thus the most comfortable, watch possible comes from the most adventurous modern spirits, and pushes against the limits of the possible.

Interestingly enough, if we look at old watchmaking manuals, we discover that Pierre-Augustin Caron de Beaumarchais, iconic figure of the Enlightenment age, and his brother-in-law, Jean-Antoine Lépine, were already making inroads in this direction in the 18th century, developing a movement characterized by a single mainplate and equipped with a cylinder escapement, which was much thinner than the traditional fusée-chain system. The fight against watchmaking "obesity" had begun. A merciless struggle that led to the present-day creation of components of a hair's thickness, with measurements precise to a micron (0.001mm), with the awareness that such mechanisms become more fragile as they get thinner.

In the year 2013 alone, we saw how invested watchmakers are in the contest to create the "thinnest watch in the world," without openly declaring their involvement in a competition for the trophy. The field of manual-winding watches shows great evidence of this! After the Vacheron Constantin Historique Ultra-fine 1955 and its 4.13mm of thickness, including the case, Jaeger-LeCoultre celebrated its 180th anniversary by presenting its Master Ultra Thin Jubilee at SIHH 2013, a watch that boasts a thickness of 4.05mm. Have we declared a winner? That is still entirely up in the air. Before SIHH 2014, Piaget released its Altiplano 38mm 900P, which uses its caseback as a bottom plate to achieve a thinness of 3.65mm. At this level of slenderness, we get closer and closer to superlatives in the realm of miniaturization. At this level of mastery of every aspect of measuring time, we enter into an horological complication that can perhaps be best expressed by "simplicity."

Michel Jeannot

CHRISTOPHE CLARET

POKER

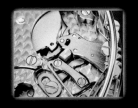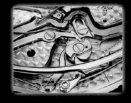

www.christopheclaret.com

GUY ELLIA

PARIS
16 PLACE VENDÔME

« REPETITION MINUTES ZEPHYR »

Premier boîtier au monde en saphir.
C'est à travers ce boîtier convexe en
saphir taillé dans la masse à très forte
résistance et exceptionnelle résonance,
que l'on découvre toute la technicité de
cette pièce de grande exception. Son
mouvement exclusif GEC 88, développé
par la manufacture Claret permet de
sonner à la demande les heures, les
quarts et les minutes sur un timbre
cathédral. A ces grandes complications,
s'ajoutent l'indication de cinq fuseaux
horaires et la réserve de marche.
Worldwide first sapphire case. It's
through a sapphire crystal case with an
impressive strength and high-resonance
that is enhanced the technical nature of
this exceptional timepiece. Its' exclusive
movement GEC 88, created by the swiss
manufacture Claret, chimes on a
cathedral tone upon request on the
hours, quarters and minutes. Added to
these complications, you can read five
different time zones and the power
reserve.

◗arije

PARIS
50 RUE PIERRE CHARRON   +33 (0)1 47 20 72 40
30 AVENUE GEORGE V   +33 (0)1 49 52 98 88
3   RUE DE CASTIGLIONE   +33 (0)1 42 60 37 77

CANNES
50 BOULEVARD DE LA CROISETTE +33 (0)4 93 68 47 73

SAINT-JEAN-CAP-FERRAT
GRAND-HOTEL DU CAP-FERRAT +33 (0)4 93 76 50 24

LONDON
165 SLOANE STREET +44 (0)20 7752 0246

One step inside any of Arije's six boutiques and one immediately thinks of Baudelaire's famous "Invitation au voyage": *Là, tout n'est qu'ordre et beauté / Luxe, calme et volupté*. At Arije, this approach permeates and guides every aspect of the experience. Whether the walls are trimmed with marble accents, or a uniform taupe, whether the lighting comes from chandeliers or modern spherical light fixtures, whether one is searching for a complicated watch or a statement piece of jewelry, Arije's boutiques offer a window onto a world where everything is exactly as it should be. There, all is order and beauty, luxury, peace and pleasure…

Serving a highly elite clientele, Arije deploys a sophisticated, perfectly multilingual staff that anticipates your needs before you are aware of them yourself. Always *au courant* of nascent trends, the boutique-salons stage a selection of watches and jewelry as daring as it is prestigious, joining centuries-old tradition to groundbreaking modernity in style and technique. Because of Arije's close relationship with manufactures and its reputation as a place for the most exigent connoisseurs, the boutiques are often the first in the world to offer new models and the only spots to procure certain exceptional limited editions.

1. **CLASSIQUE TOURBILLON EXTRA-THIN AUTOMATIC**
(Breguet)

2. **OYSTER PERPETUAL COSMOGRAPH DAYTONA IN PLATINUM**
(Rolex)

3. **PASHA 42MM SKELETON PANTHERE**
(Cartier)

4. **PATRIMONY TRADITIONNELLE CHRONOGRAPH PERPETUAL CALENDAR**
(Vacheron Constantin)

5. **JUMBO HEURE UNIVERSELLE**
(Guy Ellia)

First established in Paris's Golden Triangle in 1980, Arije has become an irresistible force in the world of luxury watches, with new locations that coincide exactly with its growing profile and desire to fulfill growing demand in France and abroad. A store on London's Sloane Street speaks to the international reach and understanding of the brand, and Cannes and Cap-Ferrat also possess locations that service Arije's demanding, fiercely loyal clientele. With roots in Paris, it should perhaps come as no surprise that in addition to its boutiques on avenue Georges V and rue Pierre Charron, Arije is opening a third location in the City of Light, on rue de Castiglione in the first arrondissement, between the Tuileries and the Place Vendôme.

A dreamlike, almost magical atmosphere pervades Arije's boutique-salons, an enchanting tapestry woven by the vision of Paris's lady of haute horology: Carla Chalouhi, who needs no last name in these rarefied circles.

Though each location expresses an individual identity and particular mission, an overarching ethos provides a guiding light. As Baudelaire might have put it, *Tout y parlerait / À l'âme en secret / Sa douce langue natale*. Everything at Arije speaks of this approach, reaching the secret corners of one's soul in a language that is universally understood: *Luxe, calme et volupté*.

6. **EXCALIBUR QUATUOR**
(Roger Dubuis)

7. **INGENIEUR CONSTANT-FORCE TOURBILLON**
(IWC)

8. **CONSTANT ESCAPEMENT L.M. ("AIGUILLE D'OR" AT THE GRAND PRIX D'HORLOGERIE DE GENEVE)**
(Girard-Perregaux)

9. **OPENWORKED ROYAL OAK OFFSHORE GRANDE COMPLICATION**
(Audemars Piguet)

9.

de GRISOGONO

GENEVE

ABU DHABI · COURCHEVEL · DUBAI · GENEVA · GSTAAD · KUWAIT · LONDON · MIAMI · MOSCOW
NEW YORK · PARIS · PORTO CERVO · ROME · Sᵀ BARTHELEMY · Sᵀ MORITZ

www.degrisogono.com

Instrumento NºUno XL

GRAND COMPLICATIONS®

THE ORIGINAL ANNUAL OF THE WORLD'S WATCH COMPLICATIONS AND MANUFACTURERS®

TOURBILLON INTERNATIONAL
A MODERN LUXURY COMPANY
ADMINISTRATION, ADVERTISING SALES, EDITORIAL, BOOK SALES

11 West 25th Street, 8th Floor • New York, NY 10010
Tel: +1 (212) 627-7732   Fax: +1 (312) 274-8418

Caroline Childers
PUBLISHER

Michel Jeannot
EDITOR IN CHIEF

EDITORS	Samson Crouhy
	Elise Nussbaum
	Julie Singer
JUNIOR ASSISTANT EDITOR	Amber Ruiz
ART DIRECTOR	Mutsumi Hyuga
CONTRIBUTING EDITOR	Fabrice Eschmann
TRANSLATIONS	Susan Jacquet
COORDINATION	Julie Mégevand
	Andrea Rohrer
VICE PRESIDENT OF OPERATIONS	Sean Bertram
DIRECTOR OF PRODUCTION	Erin Quinn
PRODUCTION COORDINATORS	Tim Maxwell
	Kari Grota Compean
DIRECTOR OF DISTRIBUTION	Mike Petre
DIRECTOR OF INFORMATION TECHNOLOGY	Scott Brookman
SALES ADMINISTRATOR	Ralph Gago

WEB DISTRIBUTION
www.modernluxury.com/watches

PHOTOGRAPHY
Photographic Archives
Property of Tourbillon International, a Modern Luxury Company

MODERN LUXURY

Lew Dickey
CHIEF EXECUTIVE OFFICER

Michael Dickey
PRESIDENT

EXECUTIVE VICE PRESIDENT AND CO-CHIEF OPERATING OFFICER	John Dickey
EXECUTIVE VICE PRESIDENT AND CO-CHIEF OPERATING OFFICER	Jon Pinch
CHIEF FINANCIAL OFFICER	JP Hannan
GENERAL COUNSEL	Richard Denning

CONSTANT ESCAPEMENT L.M.

THE GREATEST INVENTION SINCE THE TOURBILLON

GIRARD-PERREGAUX 09100-0002 CALIBER,
MANUAL WINDING MECHANICAL MOVEMENT
HOUR, MINUTE, CENTRAL SECOND, LINEAR POWER RESERVE INDICATOR
6-DAY POWER RESERVE - 48MM WHITE GOLD CASE WITH SAPPHIRE CRYSTAL
CASE-BACK, ALLIGATOR STRAP WITH FOLDING BUCKLE

GPHG
GRAND PRIX D'HORLOGERIE DE GENÈVE
—— 2013 ——
"Aiguille d'Or" Grand Prix

DE BETHUNE

L'ART HORLOGER AU XXIᴱ SIÈCLE

D B 2 8   S K Y B R I D G E

EXPLORING INFINITY

Westime

From international timepiece collectors, to customers who are considering their first purchase of a watch, Westime's three boutiques in Southern California cater to them all by offering a broad and deep selection of today's most desirable watches.

For a quarter of a century, Westime has distinguished itself as the ultimate retail destination specializing in extraordinary watches. John Simonian, a third-generation watch expert with a passion for mechanical timepieces, founded Westime in 1987 when he opened the first boutique on Los Angeles's West Side. From its earliest days, Westime catered to a clientele that ranged from Hollywood celebrities and professional athletes, to the region's influential residents and international business travelers. Westime has since earned the return business of discriminating clients from around the globe.

Today, Westime's three locations reflect the iconic styles of their surrounding neighborhoods. Westime Beverly Hills is an intimate, multi-level boutique at the heart of the city's most glamorous shopping district. The boutique is located on Via Rodeo, a European-style shopping promenade that has emerged as a global destination for the finest watch, jewelry and accessories brands. High ceilings, marble flooring, contemporary furnishings, and showcases that feature champagne water-relief woodwork complement a precisely edited selection of limited editions, high complications and even custom timepieces created exclusively for Westime's clientele. Every member of Westime's affable, multi-lingual staff is dedicated to providing exceptional service—from explaining the specifics of complications, to hand-delivering a watch across the country.

Westime La Jolla is located north of San Diego in one of the country's most beautiful seaside communities. The light-filled store resides among charming shops, galleries and cafes on elegant Prospect Street, just steps from the Pacific Ocean. The boutique's gray slate flooring, natural wood and glass watch cases and open floor plan invite customers to browse casually. The experienced Westime staff provides such special services as watch repairs and water resistance tests.

Westime Sunset is the newest and largest boutique in the family. Its location on Sunset Boulevard in West Hollywood places it among the chic restaurants and boutiques of Sunset Plaza and the legendary nightclubs of the Sunset Strip. The 6,600-square-foot Westime Sunset store reflects the bold buildings and signage of the neighborhood. Perforated and backlit metal panels wrap the asymmetrical façade, while a front wall of windows allows passersby to see the brightly lit scene inside. Asymmetrical angles and high-contrast materials including glass, Venetian plaster, steel, walnut and polished concrete create a gallery-like setting inside the two-story space. Custom corners for Audemars Piguet, Breitling, Omega and Buben & Zorweg enhance the shopping experience for fans of those popular brands.

◄ **CLASSIC FUSION CLASSICO ULTRA-THIN**
(Hublot)

Westime is frequently noted as one of a dozen multi-brand retailers in the world that influences trends in the watchmaking industry. Led by John Simonian, his son Greg and daughter Jennifer, the company is dedicated to offering the most important creations from traditional watch brands, while also promoting the new guard in haute horology. Recent pieces available exclusively at Westime include the limited edition MusicMachine by Reuge and MB&F in Westime blue; Dior's Grand Bal Piece Unique; and Bulgari's Il Giocatore Veneziano one-of-a-kind minute repeater. At Westime's boutiques, there is always something new to discover.

Westime also operates the Richard Mille Boutique Beverly Hills and Hublot Beverly Hills. The company is proud to support numerous charitable causes including After-School All-Stars, Scripps and

BRANDS CARRIED

AARON BASHA	H. MOSER & CIE.
ALPINA	HUBLOT
AUDEMARS PIGUET	IKEPOD
BELL & ROSS	LONGINES
BLANCPAIN	LOUIS MOINET
BREITLING	LUDOVIC BALLOUARD
BRM	MAÎTRES DU TEMPS
BUBEN & ZÖRWEG	MB&F
BULGARI	MONTEGRAPPA
CARRERA Y CARRERA	NIXON
CHOPARD	OMEGA
CUERVO Y SOBRINOS	PIERRE KUNZ
DE BETHUNE	REUGE
DEVON	RICHARD MILLE
DEWITT	ROGUE DZN
DIOR	RUDIS SYLVA
DÖTTLING	SHAMBALLA JEWELS
ERNST BENZ	SNYPER
FRANCK MULLER	STRUT
FREDERIQUE CONSTANT	TAG HEUER
GHISO	TISSOT
GIRARD-PERREGAUX	UGO CALA
GIULIANO MAZZUOLI	ULYSSE NARDIN
GLASHÜTTE ORIGINAL	URWERK
GREUBEL FORSEY	VERTU
HAMILTON	VISCONTI
HARRY WINSTON	ZENITH
HAUTLENCE	

▲ **MUSICMACHINE**
(Reuge and MB&F)

BEVERLY HILLS
254 North Rodeo Drive • Beverly Hills, CA 90210
Tel 310-271-0000 • Fax 310-271-3091

LA JOLLA
1227 Prospect Street • La Jolla, CA 92037
Tel 858-459-2222 • Fax 858-459-1234

WEST HOLLYWOOD
8569 West Sunset Blvd. • West Hollywood, CA 90069
Tel 310-289-0808 • Fax 310-289-0809

www.westime.com

GRAND BAL PIECE UNIQUE
(Dior)

IL GIOCATORE VENEZIANO
(Bulgari)

RM 11-01
(Richard Mille)

BR MINUTEUR TOURBILLON TITANIUM · Ldt 30 pcs · 44 X 50mm · Mechanical manual wind tourbillon · Minuteur timer with flyback function
3 Days power reserve indicator · Titanium finish case · Carbon fiber dial · Bell & ross Inc. +1.888.307.7887 · information@bellrossusa.com
e-Boutique: www.bellross.com

Summary

DESIGN AND **TECHNOLOGY.**

LUMINOR 1950 CHRONO MONOPULSANTE
8 DAYS GMT TITANIO (REF. 311)
AVAILABLE IN STEEL, TITANIUM AND CERAMIC

ASPEN · BAL HARBOUR SHOPS · BEVERLY HILLS · BOCA RATON · DALLAS
FORUM SHOPS AT CAESARS · LA JOLLA · NAPLES · NEW YORK · PALM BEACH

Exclusively at Panerai boutiques and select authorized watch specialists.

ACADEMY CHRISTOPHE COLOMB HURRICANE

The world's only watch to associate the gyroscopic "Gravity Control"
module and the fusée and chain mechanism, this model with its 939 parts
stands out as the most imperturbable timepiece of them all. Born from the
legendary expertise of the Manufacture Zenith, it conveys the enterprising
spirit of the first conquerors, the audacity that even defies the forces
of nature.

ZENITH
SWISS WATCH MANUFACTURE
SINCE 1865

WWW.ZENITH-WATCHES.COM

Web Site Directory

A. LANGE & SÖHNE	www.alange-soehne.com		**LONGINES**	www.longines.com
AUDEMARS PIGUET	www.audemarspiguet.com		**PANERAI**	www.panerai.com
BELL & ROSS	www.bellross.com		**PARMIGIANI**	www.parmigiani.ch
BLANCPAIN	www.blancpain.com		**PATEK PHILIPPE**	www.patek.com
BREGUET	www.breguet.com		**PIAGET**	www.piaget.com
CHANEL	www.chanel.com		**RICHARD MILLE**	www.richardmille.com
CHAUMET	www.chaumet.com		**ROGER DUBUIS**	www.rogerdubuis.com
CHRISTOPHE CLARET	www.claret.ch		**STUHRLING ORIGINAL**	www.stuhrling.com
DE BETHUNE	www.debethune.ch		**TAG HEUER**	www.tagheuer.com
DE GRISOGONO	www.degrisogono.com		**VACHERON CONSTANTIN**	www.vacheron-constantin.ch
DIOR	www.dior.com		**VULCAIN**	www.vulcain-watches.com
F.P. JOURNE	www.fpjourne.com		**ZENITH**	www.zenith-watches.com
FRANCK MULLER	www.franckmuller.com			
FRÉDÉRIQUE CONSTANT SA	www.frederique-constant.com			
GLASHÜTTE ORIGINAL	www.glashuette-original.com		**RELATED SITES**	
GIRARD-PERREGAUX	www.girard-perregaux.com		**BASELWORLD**	www.baselworld.com
GUY ELLIA	www.guyellia.com		**SIHH**	www.sihh.org
HUBLOT	www.hublot.com			
IWC	www.iwc.com		**AUCTION HOUSES**	
JACOB & CO.	www.jacobandco.com		**CHRISTIE'S**	www.christies.com
JAEGER-LeCOULTRE	www.jaeger-lecoultre.com		**SOTHEBY'S**	www.sothebys.com

THE ULTIMATE IN PRECISION TIMING

World First! TAG Heuer introduces the first ever column wheel integrated mechanical chronograph displaying the 1/100th of a second with a striking central hand allowing an easy reading. The Heuer Carrera Mikrograph hits 360,000 beats per hour - the ultimate in precision timing.

Carrera Mikrograph

 TAGHeuer

SWISS AVANT-GARDE SINCE 1860

Index

IWC PORTUGUESE.
ENGINEERED FOR NAVIGATORS.

——— **Portuguese Perpetual Calendar.
Ref. 5023:** One thing at IWC always remains the same: the desire to get even better. Here is one of the finest examples, with the largest automatic movement manufactured by IWC, Pellaton winding and a seven-day power reserve. The perpetual calendar shows the date and moon phase, and the

year – until 2499 – is shown in four digits. In short: a watch that has already written the future. **IWC. ENGINEERED FOR MEN.**

Mechanical IWC-manufactured movement, Pellaton automatic winding system, 7-day power reserve with display, Perpetual calendar (figure), Perpetual moon phase display, Antireflective sapphire glass,

Sapphire-glass back cover, Water-resistant 3 bar, 18 ct red gold

SCHAFFHAUSEN

Elegance is an attitude

Simon Baker
Simon Baker

LONGINES®

Conquest Classic

Elegance is an attitude

Kate Winslet

Kate Winslet

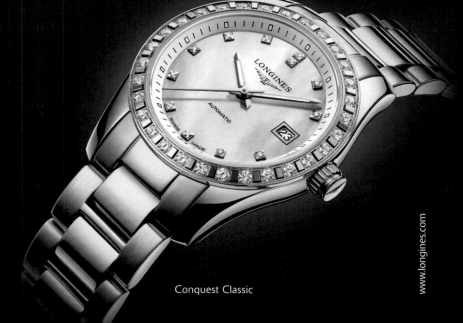

LONGINES®

Conquest Classic

Index

ROGER DUBUIS

HORLOGER GENEVOIS

EXCALIBUR
Exclusive RD01SQ Manufacture movement
Skeleton Double Flying Tourbillon

GENÈVE

**The only Manufacture to be 100% Poinçon de Genève certified.
The most demanding signature in fine watchmaking.**

1-888-RDUBUIS - www.rogerdubuis.com

ONE ADJUSTMENT EVERY 100 YEARS.
THE KIND OF COMPLICATION THAT MAKES LIFE EASIER.

Index

Index

STÜHRLING

WWW.STUHRLING.COM

DESIGNED TO

INSPIRE

The Damier Tourbillon is one of Stuhrling Original's most ornate creations. The solid stainless steel case has an adorned bezel decorated with an elegant diamond pattern. The dial has a Guilloche like pattern along with individually hand-applied shark's teeth markers. This watch is equipped with a genuine mechanical tourbillon movement that includes dual-time, animated AM/PM indicator, and an open heart that showcases the rotating tourbillon cage.

Greg Simonian
President of Westime

Since 1987, Westime has provided a welcoming home for haute horology's most passionate enthusiasts. Led by President Greg Simonian, the multi-brand retailer **CONTINUES TO SHOWCASE THE LIVELY SPIRIT OF LUXURY WATCHMAKING, CELEBRATING THE CURIOSITY OF ITS AVID VISITORS**.

Westime opened its doors in 1987 and has remained a family business ever since. How do you maintain this tradition while expressing your own individual personality?

Maintaining the traditions established by my parents is natural. After all, I was raised in the Westime stores so they have always been a part of my life! I also believe a successful business must evolve with the times, so I am always open to making changes, everything from store hours, to carrying new collections, to opening new boutiques in up-and-coming neighborhoods.

Given its status as a globally renowned multi-brand retailer, does Westime see itself as an educational window into the world of haute horology?

Absolutely, and it is a role we foster. We know our boutiques are the only places in the country to see certain models of watches, as well as a wide variety of mechanical timepieces at various prices. We love that visitors will come into Westime, enjoy a coffee, discover new watches that have just arrived and converse with our staff about anything from how mechanical timepieces work to fashion trends! Only at a multi-brand, brick-and-mortar store like Westime can clients educate themselves by seeing such a variety of watches and complications, side-by-side and on the wrist.

"Only at a multi-brand, brick-and-mortar store like Westime can clients educate themselves by seeing such a variety of watches and complications, side-by-side and on the wrist."

◄ **GREG SIMONIAN**

Describe the average Westime customer.

Westime's customers vary pretty widely, especially among our three locations. Westime Beverly Hills is quite popular with international travelers who stay in the city's five-star hotels, which are all within walking distance of our location on Rodeo Drive. Westime Sunset in West Hollywood is among restaurants and shops that are very popular with homeowners in the nearby Hollywood Hills, and travelers who appreciate the hip neighborhood. Westime La Jolla attracts the owners of vacation homes and many travelers.

Have you seen a change in the taste of haute horology enthusiasts over the last few years?

Since every enthusiast is so particular, I wouldn't put them all in the same category. Some are just beginning to develop a passion for horology, while others have been building a collection for years, if not decades. Some only want limited editions, others only seek out models from one particular manufacturer, and others collect a specific complication across many brands. Clients' preferences are really personal!

What atmosphere do you seek to create in your stores? What do you want your customers to feel upon entering a Westime location?

The most important feeling for them to feel upon entering Westime is comfort. I want them to feel like they can browse without pressure, ask any question they like, and try on the watches. We made sure that our showcases never form a barrier between the customer and the salesperson, like you see in department stores. Instead they can wander from display to display, like they would in a gallery, and try on watches in a variety of settings within the store to suit their comfort level.

Why is it so important that your staff be multi-lingual?

Los Angeles is a global capital and its residents not only speak a multitude of languages, but so do the many international travelers who come all year long for work or pleasure. By speaking the first language of our clients, we can provide the best service.

What makes the MusicMachine such an exceptional creation?

First of all, you don't find hand-crafted music boxes on every street corner! They are exceedingly rare today, and Reuge in Switzerland is, for all practical purposes, the last manufacturer of these amazing pieces that are part artwork, and part mechanical device. The MusicMachine is the result of Reuge teaming up with independent watch brand MB&F, and the DNA of both companies is immediately visible in it! The blue limited edition is available exclusively at Westime.

What is the importance of the multi-brand retail experience in haute horology?

The most important is that we offer variety compared to a mono-brand boutique experience. Equally as important is the fact that as business owners we have the power of instant decision-making. So if a client ever wants to speak to the owner, I am right here. It is not just a sales clerk in the store. And we can follow our own instincts based on our many years in our own market, and adapt our inventory, our marketing, our decor and our events quickly to adapt to changing tastes or get in front of a trend without waiting for a directive from HQ. We're also a bastion of discovery, since we can introduce exciting independent watch brands to our clients.

What complication is most popular among Westime customers?

There is not a single complication that overshadows the others. But tourbillons remain very in-demand, and there is interest in calendar complications, chiming watches, moonphases, and even some of the more esoteric complications like G-force sensors or miniature racing ...

Do you have a favorite complication?

Any complication that's a first!

How difficult is it for you to keep up with the emergence of new brands? How do you maintain a balance between the timeless appeal of classical timepieces and the new wave of ultra-modern innovative watches?

The annual watch fairs in Geneva and Basel provide a great opportunity to see what up-and-coming watchmakers are doing. I always make time to visit with some of them. Whether Westime decides to carry their watches hinges on the appeal of the watch itself, but also the watchmaker and executives since we would all become close working partners.

Westime operates three locations, all in Southern California. Are there any plans of further expansion?

I definitely would not rule out further expansion!

You sold your first watch at the age of 12. How far have you come since that day?

We've opened five new stores since then, and moved from offering very affordable watch brands to a range that extends to watches over $1 million. It has been a period of extraordinary growth for Westime!

Retrograde and Jumping Systems, Alternative Displays

Christophe Claret
X-TREM 1

Wilhelm Schmid

CEO of A. Lange & Söhne

A tradition of perpetual progress

A. LANGE & SÖHNE EPITOMIZES A BRILLIANT SYNERGY BETWEEN INNOVATION AND TRADITION, BETWEEN THE TIMELESSNESS OF THE PAST AND AN APPETITE FOR NEW ACHIEVEMENTS. Committed to a philosophy of uncompromising craftsmanship, CEO Wilhelm Schmid guides the Saxon horologer toward the future with a tremendous respect for the finesse and refinement of the watchmaking art.

What is A. Lange & Söhne's place in the world of haute horology? In other words, who is A. Lange & Söhne?
Rooted both in the Saxon watchmaking and the German engineering tradition, A. Lange & Söhne stands for unmatched craftsmanship and technical solutions that often define benchmarks. In its 168 years of history, A. Lange & Söhne has taken the lead in the development of a German watch culture. In the last 20 years, we have explored almost every complication. With meaningful innovations we have set many milestones in fine mechanical watchmaking.

How would you describe 2013 in terms of the brand's production and sales?
2013 has been a year of superlatives for us: We launched ten new models—more than ever before—and the most complicated watch ever made by A. Lange & Söhne. We presented a multitude of interesting new products at the SIHH in Geneva and can therefore look forward to 2014 with confidence.

How has the world of high-end watchmaking changed in recent years?
In the last few years, customers have become more knowledgeable and attach more importance to watchmaking excellence, precision and craftsmanship. As a result we have seen an increasing demand for highly complicated timepieces. On the other hand, the classic three-hand watch has lost nothing of its fascination and attractiveness.

Have advances in technology allowed you to expand your production capabilities while preserving your uncompromising approach to quality?
Our company has grown continuously over the past few years because the increased complexity of our products. The high share of manual work is a key element of the brand identity. This will not change in the future. Therefore, we can only grow slowly and sustainably by increasing the number of qualified employees.

◁ Lange Zeitwerk Striking Time

How important is global presence to A. Lange & Söhne?
Very important. In the last 20 years, A. Lange & Söhne has developed a loyal following across the globe. Our watches are meanwhile available in 60 countries and the brand enjoys a good presence across all markets. Over the past ten years the brand's internationalization has proceeded and we have entered new markets in Southeast Asia, South America and Australia.

How does a manufacture as proudly traditional and classical as A. Lange & Söhne keep up with an increasingly technological and modern global culture?
It is our belief that tradition is preserved progress and progress is continued tradition. In times determined by technological acceleration and mass production there is a growing desire to slow down and an increased demand for handcrafted one-off objects. High-end mechanical timepieces fulfill this desire for deceleration because they symbolize the value of time. Winding a beautiful watch is a great way to "unwind," to relax.

▶ Wilhelm Schmid, CEO

How would you explain the virtues of mechanical haute horology to a younger generation inundated with electronic devices?
It's difficult to explain it. I would rather try to introduce them to our philosophy of fine watchmaking. People who have experienced our watchmaking, either in our manufacture or somewhere at events in the world, who have looked over the shoulders of a skilled and dedicated watchmaker, finisher or engraver, have understood the value of a handmade mechanical timepiece and how it embodies the personalities of the people who designed and crafted it.

Does one particular complication best represent the spirit of A. Lange & Söhne?
The spirit of the brand is reflected by every timepiece in our collection—from entry level to the most intricate horological masterpiece. But there is always room for highlights and records. One of them is the Grand Complication, presented in 2013. The most complicated wristwatch ever built by A. Lange & Söhne combines seven, partly very rare, complications. It marks the beginning of a new era in the history of our brand.

Can you provide a glimpse into the future of the brand?
Unfortunately, I cannot comment on future developments. Full of creative ideas and innovative energy, we will continue our policy of watchmaking excellence to answer the unrelenting demand for sophisticated timepieces that will fascinate watch lovers.

Gaetan Guillosson

President of A. Lange & Söhne North America

Form and function, an irresistible alliance

A. LANGE & SÖHNE CHALLENGES THE BOUNDARIES OF THE DIAL, NEVER FORGETTING THE PRIMORDIAL IMPORTANCE OF EFFICIENCY AND LEGIBILITY. With each newly presented complication, the brand demonstrates that sophistication is at its most beautiful when it accomplishes its truest ambition of usefulness and visual simplicity.

What is the particular set of horological values that guides A. Lange & Söhne?
The brand essence comprises three aspects:
1 – A. Lange & Söhne has a long-standing tradition of sculpting Germany's watchmaking culture.
2 – A. Lange & Söhne watches feature an unmatched level of craftsmanship and at the same time deliver meaningful contributions to fine watchmaking that very often enter uncharted territory.
3 – In our untiring quest for perfection we never compromise on quality and very often go the extra mile.

You seem to be continuously redefining the architecture of the dial in a quest for optimal legibility. Why is clarity so important to A. Lange & Söhne?
It has always been the ambition of our caliber designers to create watches that are useful and have never been made this way before. The Lange 1 Tourbillon Perpetual Calendar, which we presented at this year's SIHH, is a good example of a complication that meets both requirements. It is the first time that the months are displayed by means of a rotating

peripheral and instantaneously switching ring. The underlying design is absolutely innovative and extremely complicated. This is what makes this development so attractive, but originality is not an end in itself. This kind of display makes it possible to present an abundance of information in a superbly legible layout. It is, by the way, not for the first time that we take great engineering efforts to achieve clarity or to ease the operation of a watch and it will continue to be something we always strive to perfect.

The Zeitwerk displays the hours and minutes by way of two jumping-numerals aper-tures. Explain this daring leap from convention.
Lange's caliber engineers and product designers sought to devise a watch that would be evolutionary and progressive in every respect. Combining the principles of a mechanical watch and a modern time indication format, the first mechanical wristwatchwith a truly eloquent jumping numeral display was created. A feat that no other watchmaker has successfully achieved. It is a watch that reinterprets time in an era of change and hence deemed as "the new face of time." The Lange Zeitwerk was a bold step forward for A. Lange & Söhne. It was a new concept where you could read time at a glance. A technical and design marvel, it looks different from any other timepiece. As progressive as this watch with the A. Lange & Söhne signature may be, it remains a staunch advocate of classic horological values, truly making it a new way of looking at time.

◁ Richard Lange Terra Luna

> Gaetan Guillosson
> President of A. Lange & Söhne
> North America

< Grand Lange 1 Moon Phase

< 1815 Tourbillon

The Grand Complication combines a perpetual calendar, grande sonnerie, and a split-seconds chronograph. What does this exceptionally sophisticated triumph mean to you?

The Grand Complication marks a new era for A. Lange & Söhne. With the unique combination of seven rare complications in one wristwatch, it is not only a benchmark for A. Lange & Söhne; it is also a benchmark in haute horology. This is truly a piece that our master watchmakers have learned from and will use to continue to strive further in the ongoing quest to make the finest timepieces.

The Auf/Ab power reserve display has become an emblematic signature of the brand. How important is A. Lange & Söhne's German heritage?

As a manufactory that is deeply rooted in the Saxon watchmaking tradition, and as the biggest employer in Glashütte, we feel that that it is our duty to stay committed to our German heritage by incorporating as much of it as possible into our timepieces.

Do you draw a lot of your design inspirations from the aesthetics of vintage pocket-watches?

There are certain families that are definitely inspired by vintage pocket-watches. For instance, last year we launched three new timepieces in the 1815 family. The railway track minute scale, gold and blue hands, and beveled case are an homage to the pocket-watch.

Is understated elegance a central ambition in your creation of timepieces?

It can be said that the style of A. Lange & Söhne is characterized as contemporary classical elegance that stands the test of time.

The Richard Lange Tourbillon "Pour le Mérite" boasts a pivoting dial segment that retracts to reveal the world's first-ever tourbillon with a stop-seconds mechanism.

Explain the importance of horological innovation at A. Lange & Söhne.

A. Lange & Söhne has a long-standing history of setting milestones and had already created a style of its own in the mid-19th century. Above all, exciting and innovative technical features are of utmost importance for Lange. There are many examples of the Saxon ingenuity of our master watchmakers, like the outsize date, the peripheral month ring, the 31-day power reserve, the fusée-and-chain power transmission or the tourbillon stop mechanism.

BLANCPAIN

A close-up look a

"Clockwise," "counter-clockwise," "anti-
clockwise," "at 5 on the dial," "in a sweep
of the dial": we are so accustomed to the
perpetually circular dance of watch hands
that we sometimes forget that other ways
of display-ing the time have existed, do
exist and could indeed exist!

Such is notably the case with "retrograde" systems featuring hands describing
arcs of a circle before returning to
their initial position, as well as jumping
systems featuring numerals appearing
through apertures. With the renaissance
of mechanical watchmaking in the 1980s,
these two traditional display modes

alternative time displays

returned to the forefront and have often been combined to create original dials enabling timepieces to adopt an infinite variety of faces.

In recent years, however, we have been witnessing another phenomenon with the emergence of brand-new ways to show and read the time. The revolution is in progress and bids farewell to the inevitability of endlessly spinning hands. The time has now come for spectacular shows including satellite systems, pivoting cubes, linear displays, retractable or telescopic hands—and even devices incorporating liquid!

These "alternative" displays, matched by highly sophisticated mechanisms, have become a favorite playground for youthful, independent, creative talents—and a signature feature of many new brands seeking to set themselves apart from the big names of the horological mainstream.

They testify to a blend of stunning technical mastery and unbridled imagination, thereby turning some watches into magnificently playful objects.

Welcome to the world of creativity, daring, freedom and originality. Welcome to the world of new time displays!

USING HANDS SWEEPING AROUND A DIAL TO DISPLAY THE TIME IS A FAIRLY RECENT PHENOMENON ON THE SCALE OF HUMAN HISTORY, SINCE IT APPEARED ONLY AT THE END OF THE MIDDLE AGES, IN CONJUNCTION WITH THE BIRTH OF MECHANICAL HOROLOGY.

It was preceded by an extremely broad variety of methods and instruments testifying to impressive inventiveness.

The simplest and most ancient way of evaluating the time of day is of course to observe the height of the sun in the sky—or its corollary, meaning the position and length of the shadow cast on the ground by a vertical body such as a stick, a tree or a stone. To fine-tune these empirical measurements, human beings soon invented sundials or "shadow clocks," meaning instruments fitted with a fixed hand casting its moving shadow on a graduated surface that is either flat (horizontal or vertical) or rounded. Sundials were commonly used in Egypt as of the second millennium BC and initially mainly served to organize religious rites. In ancient Rome, they often adorned temples and other public buildings, while travelers and soldiers carried their portable sundials around with them. Throughout the Middle Ages, sundials remained one of the main means of getting one's temporal bearings—at least during the daytime, under clear skies.

At night, our forebears would observe stars describing arcs around the polestar, which has always been the main temporal reference. The ancient Egyptians used the rising and setting of certain stars to mark the beginning and end of the nocturnal hours.

However, our early ancestors soon realized that they could make use of certain natural phenomena such as the regular flow of water to measure durations without resorting to the sun or the stars. This resulted in the famous clepsydras or "water clocks," featuring vessels with a small hole in them, in which the level of the liquid made it possible to read off the time on a graduated vertical scale. Already used in Mesopotamia, notably for irrigation, as well as in Egypt, clepsydras were very popular in Ancient Greece. One of the best-known examples is the famous Tower of the Winds, built in the first century BC and still standing proudly with its empty reservoir at the foot of the Acropolis. Romans also made considerable use of clepsydras, which served to time, among other things, the length of defense speeches during court hearings, or to keep track of slaves' work.

Harnessing hydraulic energy also led to the development of some highly sophisticated clocks, sometimes featuring spectacular display modes. In the 2nd century BC, the famous Greek Kitesibios water clock equipped with various wheels and gears was topped by a tiny figurine pointing with a tiny rod to the hours traced on a column. The monumental Arab and Chinese clocks from the early Medieval period also used the power of water to drive various automaton systems, such as horsemen emerging from a door to mark off the hours, or tiny bronze balls falling into a vessel and causing it to vibrate.

Western countries during the Middle Ages also used to measure durations or lengths of time. Such was notably the case with "fire clocks" that took on a variety of forms: braided wicks that took a specific length of time to burn, oil lamps with graduated reservoirs, or graduated candles sometimes even featuring tiny nails that dropped to provide an audible indication! From the 14th century onwards, progress in the field of glassworking led to the creation of hourglasses, which were notably used to time sermons.

The birth of mechanical horology, in the early 14th century, did not immediately result in a new way of reading the time. The first monumental clocks, on steeples and belfries, often had no dial whatsoever and the indication of time was purely audible, frequently including the famous "jacks," figures that struck the bells.

The system based on a hand revolving around a dial—inspired by the circular displacement of the shadow on sundials—only gradually established itself during the 14th century and at the start of the 15th century. The left-to-right direction of the rotation, borrowed from horizontal solar clocks, soon became the norm, although some clocks such as that of Santa Maria del Fiore (1443) in Florence, on which the hour hand, as is the case on vertical sundials, rotates counter-clockwise.

The first pocket-watches, which appeared in the late 15th century, had only one hand, the hour hand. It would take the invention of the regulating balance-spring (1675) to popularize the use of the minutes hand, followed a century later by the seconds hand. This system, subsequently used on wristwatches, has seemed so "natural" to us ever since that it has not been superseded by quartz technology, even though the latter could perfectly well dispense with hands.

However, horologists soon began striving to develop other ways of reading time. In the 17th century, pocket-watches with retrograde displays had already made their appearance, and in the 18th century, Abraham-Louis Breguet reveled in enlivening his dials with hands of this type for indications such as the date.

The disc- and aperture-type displays that are the ancestors of our jumping systems were also featured in the first pocket-watches for long-duration indications such as the months or Zodiac signs. Their use was then extended to encompass the hours and minutes and equipped with instant-jump mechanisms. This type of "digital" indication enjoyed an unprecedented boom from around 1925 to 1935 on rectangular wristwatches in which the dial is replaced by simple apertures cut out from the flat surface, such as the "Tank à guichet" from Cartier (1928).

The late 20th-century rebirth of mechanical horology brought these alternative display modes (which vary appearances and styles) back into favor, as well as nurturing the emergence of all-new ways of indicating and reading the time.

HOW DOES IT WORK?

RETROGRADE DISPLAYS

On a retrograde display, the hand traces an arc before returning in the blink of an eye at its original position to begin its path again. The construction of the base movement is the same as for an ordinary watch, and the difference lies in the display mechanism placed just below the dial. A complex system of springs, racks and pawls transforms the circular movement of the hand into a back-and-forth movement with ultra-fast return and immediate restart. The more retrograde displays are featured on a watch, the more precision and expertise are required to build it.

JUMPING DISPLAYS

On jumping mechanisms, the hand is replaced with a disc that "jumps" instantly with each change of unit. The base movement has the same construction as that of an ordinary watch, but a system of drawback and jumper springs serves to temporarily halt the indicator, accumulating energy until it is time to release it all at once to advance to the next increment. The jumping display is generally used for units lasting a significant amount of time, such as the hour or the date. It requires excellent control over energy, especially when several mechanisms of this type must spring into action at the same time.

Retrograde Displays

Retrograde displays are very fashionable at the moment, and their use is no longer limited to the minute and the date, instead spreading to indications such as the hour, seconds, day, chronograph counter, second time zone and other functions. The balletic movement of a hand springing back to its initial position often lends retrograde timepieces a highly spectacular appearances—especially when the dial combines two, three or even four of these indications, or when the watchmakers have a little fun by complicating the display even further, or replacing hands by tiny automatons. Another popular trend opens up the heart of the mechanism, allowing the wearer to peek "backstage."

> **Van Cleef & Arpels**, *Lady Arpels Ballerine enchantée*
The Lady Arpels Ballerine enchantée by Van Cleef & Arpels is adorned with a delicate feminine figure that is half-dancer, half-butterfly, sculpted in gold and set with diamonds. Pressing the pushbutton at 8 o'clock lifts the two skirts of the tutu in turn so as to indicate the hours and minutes.

< **Van Cleef & Arpels**, *Le Pont des Amoureux*
Van Cleef & Arpels poetically reinvents the retrograde display on its model Le Pont des Amoureux, with a young woman's parasol marking the hours to the left, and her suitor's rose ticking off the minutes at right. The young lovers kiss twice a day during this continuous waltz.

⌃ Milus, *Tirion Triretrograde*

The Tirion Triretrograde by Milus features a retrograde seconds display based on three 20-second segments. The first hand moves across its arc of a circle before moving instantly back to its starting point, while the second hand takes over, and then passes the baton on to the third.

⌃ Jaquet Droz, *Date Astrale*

A specialist in original displays, Jaquet Droz has created for its Date Astrale model a retrograde date indication in which the hand is replaced by a diamond sweeping across a segment running from 10 o'clock to 5 o'clock.

⌃ Piaget
Emperador Coussin Perpetual Calendar

Equipped with an ultra-slim in-house movement, Piaget's Emperador Coussin Perpetual Calendar displays the day and the date using retrograde pointers on two fan-shaped segments respectively placed at 9 and 3 o'clock.

⌄ Maurice Lacroix, *Starside Sparkling Date*

On its Starside Sparkling Date watch, Maurice Lacroix has opted to indicate the date by a diamond sliding along a wavy line crossing the dial from 9 o'clock to 4:30, before returning in a flash to its initial position at the end of each month.

⌄ Longines, *The Master Collection Retrograde Moon Phases*

Equipped with an exclusive movement, The Master Collection Retrograde Moon Phases from Longines reveals no fewer than four retrograde displays—seconds, day date, 24-hour second time zone—on a barleycorn guilloché dial.

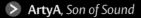 **ArtyA**, *Son of Sound*

The "Son of Sound" watch by ArtyA features a guitar-neck case and displays retrograde hours and minutes by means of hands fitted on metronome-type discs. The musical touch is further accentuated by pushers shaped like tuning pegs.

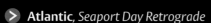 **Pierre DeRoche**, *TNT Royal Retro Power Reserve*

On the TNT Royal Retro Power Reserve from Pierre DeRoche, equipped with an exclusive Dubois Dépraz mechanism, the seconds are indicated via a succession of four retrograde displays activated via a system of gears and springs that is visible on the dial. The two remaining displays provide space for a power reserve indication.

Atlantic, *Seaport Day Retrograde*

The quartz model Seaport Day Retrograde from Atlantic displays the day with a hand that traverses an arc between 8 and 12 o'clock—completed by a double aperture for the large date at 1 o'clock and an oversized seconds subdial at 6 o'clock.

⌃ **Vacheron Constantin**,
Patrimony Contemporaine day-date bi-retrograde

The highly original Patrimony Contemporaine day-date bi-retrograde from Vacheron Constantin indicates the date on an arc from 9 to 3 o'clock, which is echoed by a second segment for the day from 7 to 5 o'clock.

⌃ **Jean d'Eve**, *Sectora II Automatic*

The Sectora II Automatic from Jean d'Eve displays the hour and minute across a 120° arc, housed in an asymmetrical case with a highly original shape.

⌃ **Pierre Kunz**
Seconde Virevoltante Rétrograde

On its Seconde Virevoltante Rétrograde, Pierre Kunz offers an amazing show of aerial acrobatics with a hand dipping and swooping around a 470° loop before returning to its initial position in 5/100ths of a second and starting all over again.

⌃ **Pierre DeRoche**
GrandCliff Double Retrograde Skyscrapers

Inspired by ever more daring, ever more vertiginous skyscrapers, Pierre DeRoche created the GrandCliff Double Retrograde Skyscrapers, with two vertical retrograde displays for the hours and the minutes.

⌃ **Cuervo y Sobrinos**
Historiador Retrogrado

The Historiador Retrogrado model from the Cuban-Swiss brand Cuervo y Sobrinos displays the day and date by retrograde indicators respectively placed at 9 o'clock and 3 o'clock—framed by a case inspired by an historical model.

⌃ **Blancpain**, *Villeret*

Upon its delicate guilloché dial, covered with translucent blue lacquer, this Villeret model from Blancpain displays a serpentine central date hand, as well as a retrograde seconds display on a 30-second segment at 6 o'clock.

Jumping Displays

In jumping displays, the indication is provided by a disc on which are written numerals that "jump" by one increment at each change—as the date does on instantaneous calendars. This system is most often used for the hour display. However, horologers are increasingly intrigued by "digital" jumping indications, which they also use for other units (minutes, seconds, date, day or month), often revealing the discs upon the dials.

Montblanc, *Nicolas Rieussec Rising Hours*

The Nicolas Rieussec Rising Hours watch by Montblanc stands out for its jumping hour system equipped with an original day/night mechanism. The hours appear in a light blue shade from 6 pm to 6 am and in black during the day.

F.P. Journe, *Vagabondage II*

On the Vagabondage II from F. P. Journe, the only indication with a hand is the small seconds counter at 6 o'clock. Three discs appearing through a mostly openworked dial indicate the hours (at 12 o'clock) and the minutes (in the center) in two bisected apertures.

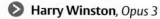

> **Harry Winston**, *Opus 3*

For the Opus 3 by Harry Winston, Vianney Halter created an all-digital display with jumping hours, minutes, seconds and date appearing in six separate "portholes," and ten independently rotating discs driven by two mechanisms totaling 250 parts.

> **De Bethune**, *Digitale*

The Digitale from De Bethune presents an entirely disc-based display, with an upper aperture for day, date and month, one in the center for the minutes and an aperture in the lower half of the dial for the hours.

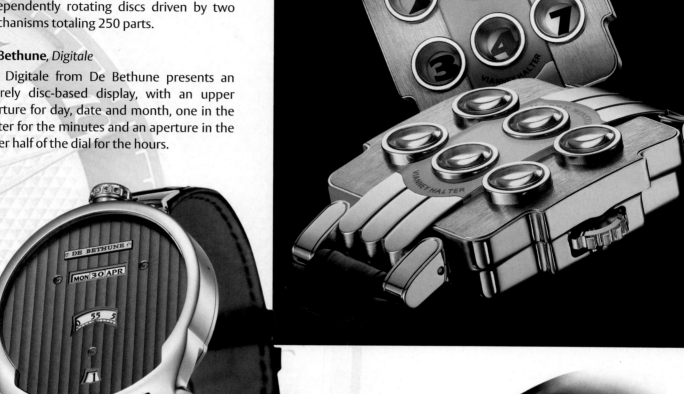

> **A. Lange & Söhne**, *Lange Zeitwerk*

The Lange Zeitwerk from A. Lange & Söhne associates traditional and digital displays, with a jumping hour on a double disc at 9 o'clock, jumping minutes at 3 o'clock and two counters with conventional hands for the small seconds and power reserve.

Fortis, *B-47 Mysterious Planets*
Developed in partnership with designer Karsten Krebs, the Fortis B-47 Mysterious Planets displays the jumping hours through five round apertures symbolizing the position of the planets around the Sun. It is available in a 500-piece limited edition.

Thomas Prescher, *Nemo Captain*

A tribute to 20,000 Leagues Under the Sea by Jules Verne, the Nemo Captain model from independent watchmaker Thomas Prescher features a triple-axis flying tourbillon appearing through a porthole at 11 o'clock along with a jumping hour display at 4:30. The minutes are indicated by the rotation of the outer tourbillon carriage.

Hysek
IO Jumping Hour

Outfitted with an exclusive movement and an atypical look, the IO Jumping Hour from Hysek presents a multi-level dial with jumping hours, dragging minutes displayed on a disc and day/night indicator.

Vincent Calabrese, *Baladin*

Vincent Calabrese, on his Baladin, proposes an unusual way of reading the time by displaying the jumping hour in a small round window that marks off the minutes by rotating around the dial.

Cartier
Rotonde de Cartier
Jumping Hours

A watch without hands, the Rotonde de Cartier Jumping Hours associates an aperture with jumping hours at 12 o'clock with a dragging minutes system equipped with a rotating disc. The hours are displayed in Roman or Arabic numerals.

Retrograde/Jumping Combinations

The jumping hour is often partnered with a retrograde minute, or other similar indications (second, date, chronograph counters, etc.). This offers a fascinating spectacle on the hour and at midnight, when the numerals jump in their apertures and the hands leap to their starting position.

> **Bulgari**, *Commedia dell'Arte*

On the Bulgari Commedia dell'Arte, the jumping hour appears through a window at 6 o'clock, while Pulcinella, Harlequin or Brighella point to the retrograde minutes with their right arm. The animation of the automatons is associated with a minute repeater system graced with cathedral-type gongs. Limited series of eight watches for each of the three versions.

< **Hautlence**, *Avant-garde HL^{RQ}03*

The Avant-Garde HL^{RQ}03 model from Hautlence associates a jumping hour at 9 o'clock and a retrograde minute at 3 o'clock, linked by an exclusive connecting-rod/crank system—all housed within an open-worked architecture that allows for beautiful 3D effects.

MB&F, *Horological Machine N° 2*

The Horological Machine N° 2 by MB&F stands out for its original two-dial look. The right-hand dial displays jumping hours and concentric retrograde minutes, while the left-hand provides retrograde date as well as twin-hemisphere moonphase displays.

John Isaac, *Jumping Hour*

Developed in collaboration with Dubois Dépraz, the limited edition Jumping Hour from John Isaac stands out for its original design, which features an elliptical case framing a round subdial with jumping hours and retrograde minutes.

Fonderie 47, *Inversion Principle*

Featuring jumping hours, retrograde minutes, a central flying tourbillon rotating in three minutes and a six-day power reserve, the Inversion Principale by Fonderie 47 is also designed as a "charity watch" to raise funds for disarmament in Africa. It is available in a 20-piece limited edition.

Bulgari

Octo Chronograph Quadri-Retro
Gérald Genta Collection

A true animated ballet, the Octo Chronograph Quadri-Retro, from Bulgari's Gérald Genta Collection, associates a jumping hour aperture with four retrograde indications: central minutes, date at 6 o'clock and chronograph hour and minute counters at 3 and 9 o'clock, respectively.

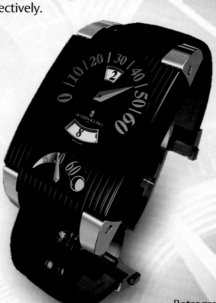

de Grisogono, *FG One*

The FG One from de Grisogono boasts two dials, one with retrograde minute on a 230° arc, instantaneous jumping hour and dragging supplementary time zone, the other with a retrograde seconds complemented by a day/night aperture linked to the second time zone.

Spero Lucem, *La Jonction3*

Paying tribute to a district of Geneva, the La Jonction model from the youthful Spero Lucem brand has a pure, restrained dial associating a jumping hour window at 11 o'clock with a retrograde minutes display at 1:30 and a flying tourbillon at 6 o'clock.

Dials go wild

Hands stopping at will, turning backwards or slowing down; hours and minutes indicated by tiny precious insects; pivoting numerals, recomposed hour markers, square wheels; watches with no time indication whatsoever... some watchmakers turn dials into a truly lively show in which visual attraction supersedes time read-off.

> **Chanel**, *J12 Rétrograde Mystérieuse*
> The central minutes hand of Chanel's J12 Rétrograde Mystérieuse runs in a traditional manner for the first ten minutes, before bumping up against the retractable crown appearing on the dial at 3 o'clock. This sets off a retrograde motion that takes another ten minutes to bring the hand to the other side of the crown, on the 20th minute.

⌃ MeisterSinger, *N°1*

The N° 1 model by German brand MeisterSinger revives the tradition of early pocket-watches by adopting a dial with the hour hand only. Minutes are read off in a more approximate mode thanks to a scale with five-minute graduations.

⌃ Hermès, *Arceau Le temps suspendu*

On the "Arceau Le temps suspendu" watch by Hermès, a pusher serves to bring the hour and minute hands to a halt at about 12 o'clock, while a further press sets them running again as usual. Meanwhile, a mischievous hand whirls continually backwards on a tiny subdial with 24 graduations.

⌖ Ludovic Ballouard, *Half Time*

On this second creation from independent watchmaker Ludovic Ballouard, Half Time, the hour hand is replaced by a series of Roman numerals cut in half and thus illegible—apart from that of the hour in progress, instantaneously appearing at 12 o'clock. The minutes are shown by a retrograde hand.

⌖ Pierre Kunz, *Infinity Looping*

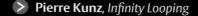

On its Infinity Looping model, the Genevan watchmaker Pierre Kunz invented a spectacular new way of reading the time. Thanks to the gravitation of a satellite pinion that moves around a fixed rack, the hours and minutes whirl around the dial, tracing a rosette shape.

⌕ Ludovic Ballouard, *Upside Down*

On the Upside Down by Ludovic Ballouard, the hour in progress is indicated by Arabic numerals the right way up, while the 11 others appear... upside down. The hour markers rotate instantly when the minute hand reaches the 12 o'clock position.

Chaumet, *Attrape-moi... si tu m'aimes*

The Attrape-moi... si tu m'aimes watch by the Parisian jeweler-watchmaker displays the hours and minutes in the form of a perpetual chase starring a spider and a bee (the brand emblem) rotating around the dial. The collection also features other insects such as the dragonfly and the butterfly.

Hermès, *Cape Cod Grandes Heures*

Hermès has equipped its Cape Cod Grandes Heures watch with an exclusive module that speeds up or slows down the hour hand, while the minutes and seconds tick by at the usual pace. A fine tribute to the relativity of time.

Maurice Lacroix
Masterpiece Régulateur Roue carrée

On the Masterpiece Régulateur Roue carrée by Maurice Lacroix, the hour appears in one of the angles of the square wheel driven by another wheel shaped like a cloverleaf—with both appearing on the dial. Available in a 99-piece limited edition.

RJ-Romain Jerome
Day & Night

The nine-piece Day & Night limited series by RJ-Romain Jerome is distinguished by its dial free of any temporal indication, other than two sequential tourbillons serving to tell day from night.

Bell & Ross, *BR Twelve O'Clock*

Issued in a 12-piece limited edition, the BR Twelve O'Clock by Bell & Ross is a highly original clock made up of 12 watches. The numerals are spread among three discs and on each hour, a single watch displays a legible number.

Linear displays

Some brands make a clean break from the circular sweep of dial hands or the segments featured on retrograde displays by offering linear horizontal or vertical displays of the hours, minutes or seconds, as well as for certain auxiliary indications (chronograph, power reserve, etc.). This approach is often coupled with the development of certain highly innovative mechanisms (cylinders, magnetic spheres, chains, belts, etc.).

> **Audemars Piguet**, *Royal Oak Carbon Concept*

On its Royal Oak Carbon Concept watch, Audemars Piguet adopts a linear display for the 30-minute chronograph counter, composed of a double vertical scale appearing on the right-hand side of the dial (with tens from 1 to 2 and units from 0 to 9), while the elapsed time is indicated by the successive transitions of the various openings from white to black.

∧ **TAG Heuer**

Grand Carrera Calibre 36 RS Caliper Concept Chronograph

In addition to providing readings accurate to the nearest 1/10th of a second, the Grand Carrera 36 RS Caliper Concept Chronograph by TAG Heuer features a linear small seconds display with a distinctly dashboard-style look at 9 o'clock.

> **Harry Winston**, *Opus 9*

The Opus 9 by Harry Winston, developed by design engineer Eric Giroud and watchmaker Jean-Marc Wiederrecht, features a linear display of the hours and minutes by means of two small brass chains set with emerald-cut diamonds, along with vivid mandarin garnets serving as hour markers.

Christophe Claret, *DualTow*

On his DualTow watch (with tourbillon and single-pusher chronograph) Christophe Claret presents an hour and minute display by notched rubber belts inspired by caterpillar-track vehicles.

Christophe Claret, *X-TREM-1*

On the X-TREM-1 from Christophe Claret, the retrograde hour and minute hands are replaced by two small steel spheres placed in two sapphire tubes on either side of the case and which move using magnetic fields.

Hysek, *Colosso*

On its spectacular Colosso watch, Hysek provides a retrograde dual-time display via a system of Archimedes screws and pinions serving to raise or lower the two arrows pointing to the 24 hours.

Maurice Lacroix
Masterpiece Seconde Mystérieuse

The Masterpiece Seconde Mystérieuse by Maurice Lacroix offers a linear reading of the seconds in alternating vertical and horizontal 15-second cycles. The seconds hand appears to be floating or virtually levitating above a light-colored subdial.

Urwerk, *UR-CC1 Black Cobra*

Inspired by the speedometers on certain 1950s cars, the UR-CC1 Black Cobra from Urwerk presents the linear display of the hours and minutes via two cylinders. Upon reaching the far end of its trajectory, the minute cylinder returns to its initial position in one tenth of a second, and this retrograde movement pushes the hour cylinder one unit forward.

Jean Dunand, *Palace*

For Jean Dunand's Palace model, Christophe Claret designed a flying tourbillon caliber with a second time zone and power reserve indication, situated in two vertical cartridges traversed by two ovals with arrows on them.

▼ RJ-Romain Jerome, *Spacecraft*

Designed by watchmakers Eric Giroud and Jean-Marc Wiederrecht, the Spacecraft by RJ-Romain Jerome boasts a striking design reminiscent of the spaceships featured in TV series from the 1970s and 80s. A disc indicates the minutes on the top of the case, while the retrograde linear jumping hour is displayed by a red cursor visible on the side of the watch.

Disc-type systems

These days, discs no longer hide beneath the dials. Orbiting like skies in the night sky, they create lively ballets, providing a stage for all manner of fanciful ideas.

Bell & Ross, *BR01-92 Heading Indicator*

Inspired by the gyroscopic compass in aircraft cockpits, the dial of the BR01-92 Heading Indicator from Bell & Ross comprises three discs. The seconds are displayed in the center pointed to by a yellow marker and the minutes on an intermediate disc at 12 o'clock, while the hours appear opposite the yellow triangle on the outer disc.

Ressence, *Type 3*

The Type 3 model by Ressence is equipped with a satellite system with various rotating disc-type displays (hours, seconds, days), while the mobile dial itself indicates the minutes. The entire system is immersed in a visibility-enhancing liquid. The date appears around the periphery.

Harry Winston, *Opus 7*

For Harry Winston, in the Opus 7, the watchmaker Andreas Strehler created an "alternating display." A first press on the crown guard turns the disc to bring the correct figure opposite the marker at 10 o'clock, a second press displays the minutes, and a third the power reserve.

RSW, *Outland*

On Outland from the brand RSW, the traditional hands make way for three concentric discs (hours, minutes and seconds) that move around a fixed point.

Roller-type systems

While still very rare, roller-type systems are sometimes used to indicate the hours and minutes, along with other temporal data. This touch of originality is in some cases associated with equally unusual designs.

Cabestan, *Trapezium*

Equipped with a vertical tourbillon and a fusee-chain device, the highly original Cabestan Trapezium in titanium displays the hours, minutes, seconds and power reserve using a system of rotating drums.

Rebellion, *T-1000 Gotham*

In addition to its imposing tank-like titanium case, and its impressive 40-day power reserve (ensured by six barrels), the T-1000 Gotham by Rebellion features a vertical indication of the hours and minutes appearing on two rollers.

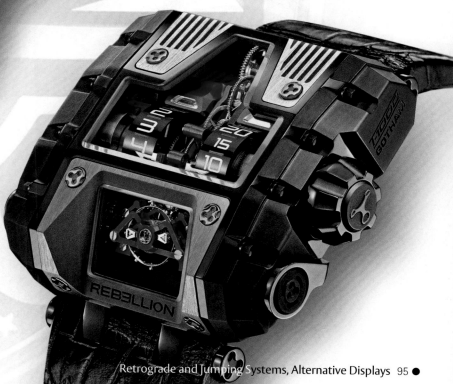

Jean Dunand, *Shabaka*

The ultra-sophisticated Shabaka model by Jean Dunand is equipped with a cathedral-gong minute repeater mechanism and is also distinguished by its instant-jump roller-type perpetual calendar indications.

3 D displays

In recent years, we have been witnessing an ever-increasing number of watches revolutionizing the indication and reading of time by means of pivoting, retractable, satellite-style or hydro-mechanical three-dimensional mechanisms—often combined with highly original case designs. These alternative systems are generally produced by the new watch brands or by independent designers and developers, sometimes working backstage for the major companies. While they do not always guarantee optimal legibility, they nonetheless display impressive inventiveness backed by peerless technical mastery.

Bulgari, *Chronograph Papillon*

On the Bulgari Chronograph Papillon from the Daniel Roth Collection, the minutes are displayed by an ingenious central device that completes a 360° rotation in two hours and boasts two independent pivoting hands that take turns traversing a 180° segment that marks 0 to 60 seconds.

MCT, *Sequential One – S110*

The Sequential One – S110 by MCT puts on a fantastic show. The hour appears in turn in one of the four display zones on the dial by means of an original system of triangular prisms closely resembling animated advertising panels. The minutes appear over a 270° arc, and at each change of hour, the mobile minute ring makes a one-quarter turn to the left to reveal the following hour.

Louis Vuitton, *Tambour Spin Time GMT*

The Tambour Spin Time GMT by Louis Vuitton marks the hours using 12 rotating rods arranged in a star shape, each tipped with a small cube. Every time the minutes move from 59 to 00, one of the cubes pivots to reveal the following hour, while the cube displaying the previous hour also makes a quarter-turn to reveal its neutral face.

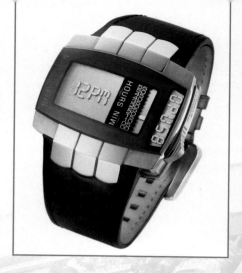

⌃ Harry Winston, *Opus 8*

For the Opus 8 by Harry Winston, Frédéric Garinaud devised a system of digital hours and minutes display inspired by pin art games. Activating a bolt on the side of the case causes the digits corresponding to the time to "rise up" from the dial, while the minutes are marked off on a vertical scale.

⌄ Harry Winston, *Opus 12*

The Opus 12 from Harry Winston, developed in conjunction with Emmanuel Bouchet, proposes an unusual way to read time, with 12 pairs of hands that pivot to mark the hours and five-minute increments. A central retrograde hand counts off the five minutes between the increments.

⌃ Harry Winston, *Opus 11*

Harry Winston's Opus 11, created by watchmaker Dennis Giguet, has no hands at all. Every 60 minutes, the large central numeral for the hour explodes into chaos before being reconstituted in a fraction of a second, using 24 mobile "pallets."

⌃ Harry Winston, *Opus 13*

Developed in collaboration with independent watchmaker Ludovic Ballouard, the Opus 13 from Harry Winston displays the time by a sophisticated system of 59 minute hands successively pivoting around the dial, along with 11 rotating hour triangles and a sliding shutter combining to stage a truly magical show.

MB&F, *Horological Machine N° 3 Frog*

Created by the MB&F watch collective, the extremely surprising Horological Machine N° 3 Frog houses the hour and minute displays on rotating bezels in its two frog's eyes.

de Grisogono
Meccanico dG, version titane noir

The Meccanico dG dual time zone watch by de Grisogono combines an analog display of the first time zone with a digital display of the second, with both driven by a completely mechanical movement comprising 651 parts. The digital second time zone indication is composed of mobile micro-segments driven by 23 cams.

MB&F, *Horological Machine N° 5 On the road again*

Paying tribute to supercars, the highly original Horological Machine N° 5 On the road again, by MB&F, features a system of louvers opening to let the light through and thus charge the SuperLumiNova of the hour and minute discs, which are placed flat on the movement and appear vertically on the side dial by means of an optical effect.

HYT, *H2*

The H2 by HYT (known as the "Hydro Mechanical Horologists") features a standout retrograde display of the hours by means of a fluorescent liquid propelled by the alternating compression/decompression motion of two mini-reservoirs with V-shaped pistons at 6 o'clock. This hydro-mechanical device is complemented by a central jumping minute hand.

Cyrus, *Klepcys*

Created by the watchmaker Jean-François Mojon, the movement of the Cyrus Klepcys model displays the hour using a retrograde hand mounted upon two rotating cubes that indicate the day and night. The minutes and seconds are read off two central discs, and the date is indicated by a retrograde hand with a three-dimensional rotating tens column.

Urwerk, *UR-103*

The UR-103 by Urwerk, a brand specializing in innovative displays, is comprised of satellites, each marking off three hours. In turning, the forward-most satellite simultaneously marks off the minutes on a curving scale.

Urwerk, *UR-202*

The UR-202 by Urwerk indicates the time by means of three rotating cubes taking turns, while a telescopic hand displays the minutes by precisely following the angular path of the graduated scale.

A. LANGE & SÖHNE

LANGE ZEITWERK – REF. 140.025

The Lange Zeitwerk is powered by the manual-winding Lange proprietary L043.1 caliber, which contains 388 components and 68 jewels. Decorated and assembled by hand, the movement beats at 18,000 vph and possesses a precision beat-adjustment system with a lateral setscrew and whiplash spring. The dark tinted sapphire crystal glass has a special UV-transparent PVD coating that permits light to enter and charge the numerals with enough photonic energy to reliably emit the time for several hours during the night. The case is platinum and the series is limited to 100 pieces.

BELL & ROSS

WW1 HEURE SAUTANTE

The WW1 Heure Sautante is a Bell & Ross original complication for those who are passionate about timekeeping. The Arabic numeral hours are displayed through a large fixed window at the top of the dial, replacing hour hands. This aperture is vertically aligned with both the power reserve indicator at the bottom of the dial and the central minute hand. The power reserve indicator is exceptionally important on this timepiece because the mechanism that triggers the jumping change in hours requires considerable energy. The wearer is thus made aware of how much power remains in the WW1 Heure Sauntante: essential information for seamless functioning of the movement.

CHANEL

J12 RÉTROGRADE MYSTÉRIEUSE – BLACK CERAMIC – REF. H2971

1. Tourbillon
2. Digital minutes display
3. Retrograde minutes hand
4. 10-day power reserve
5. Retractable vertical crown

This perfectly round watch with a 47mm diameter has been designed without a side crown to ensure optimum comfort on the wrist. The work on this complication was entrusted to one of the most "state-of-the-art" watchmaking design and construction workshops: the Giulio Papi team (APRP SA).

Limited and numbered edition of ten pieces in black high-tech ceramic and 18K white gold or ten pieces in black high-tech ceramic and 18K pink gold. The white and chromatic high-tech ceramic edition are unique pieces.

CHANEL

J12 RETROGRADE MYSTERIEUSE – WHITE CERAMIC

When technical prowess compliments the aesthetics
CHANEL RMT-10 Calibre
- R for Retrograde
- M for Mysterious
- T for Tourbillon

After introducing the first tourbillon featuring a highly innovative high-tech ceramic bottom-plate, CHANEL confirmed its status as a pioneer of contemporary watchmaking by creating the J12 Rétrograde Mystérieuse. The Rétrograde Mystérieuse is a concentrated dose of innovation, bringing together complications and a world première.

CHAUMET

ATTRAPE-MOI... SI TU M'AIMES CREATIVE COMPLICATION – REF. W16199-BC1

This seductive dance between a spider and Chaumet's iconic bee displays time with an imagination only matched by unprecedented mechanical innovation. Thanks to two off-centered rails on a brilliant-cut-diamond-set mother-of-pearl dial, the CP12V-XII self-winding caliber drives a display of time that eschews hands and conventional circularity. Pursuing independent paths upon an exquisitely crafted web, the pink-gold spider and diamond-studded bee indicate the hours and minutes, respectively. Framed by a bezel set with diamond indexes for the minutes and 12 pink-gold cabochons for the hours, the two insects act out on a wily game. A white leather strap set with two rows of diamonds ensures a luxurious fit around the wearer's wrist.

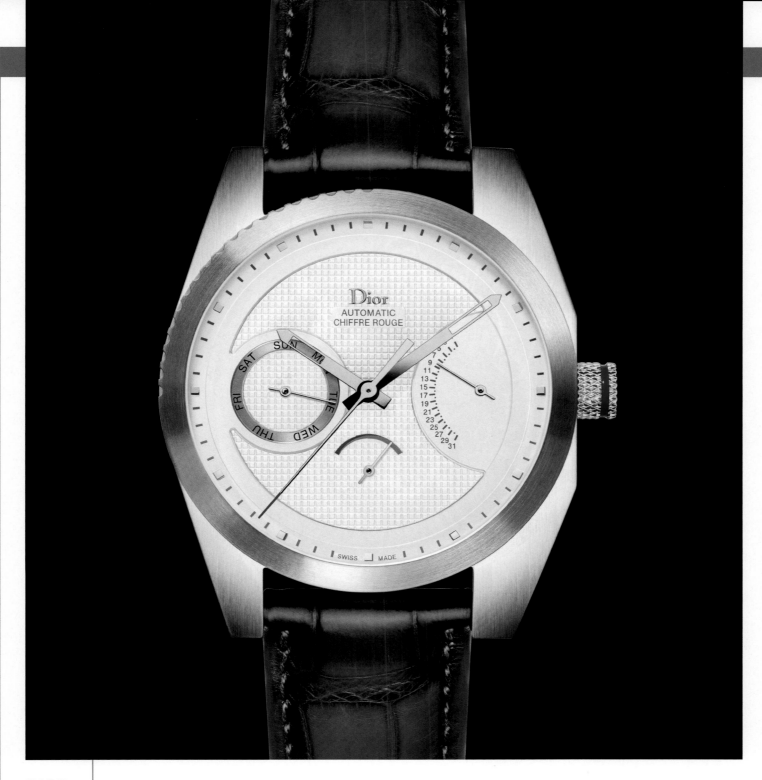

DIOR

CHIFFRE ROUGE C01 – REF. CD084C10A001

A new timepiece with a retrograde display completes the Chiffre Rouge collection and is the quintessential watch for the formal and stylish man. The delicate balance of the retrograde date display, day counter and power reserve indicator makes this model elegant in design while complex in mechanics. The Chiffre Rouge C01 timepiece is embellished with a cotton piqué patterned dial and sits gracefully on a hand-sewn alligator strap. It is also available in two versions: a silver opaline dial on a slate gray strap and a black dial on a black strap, both available in limited editions of 200 pieces each.

JACOB & CO.

PALATIAL FLYING TOURBILLON JUMP HOUR RETROGRADE MINUTES
REF. 150.510.24.NS.PB.1NS

Complete with jumping hours with an instantaneous-jump disk at 10:00 along with a retrograde minutes sector between 12:00 and 3:00, the Palatial Flying Tourbillon Jump Hour Retrograde Minutes is powered by a hand-wound JCBM02 caliber. With a power reserve of 100 hours, this model oscillates at 21,600 vph. The titanium case, at a diameter of 43mm is topped off with a glimmering 5N 18K gold crown. The vibrant sea-blue hue of the mineral crystal dial contrasts from the gilted lettering and titanium tourbillon within. The power reserve is indicated on the caseback, where the movement's intricate finishes may be admired through a scratch-resistant sapphire crystal.

VACHERON CONSTANTIN

PATRIMONY CONTEMPORAINE RETROGRADE DAY AND DATE – REF. 86020/000G-9508

Powered by the mechanical automatic-winding Vacheron Constantin Caliber 2460 R31 R7, this Patrimony Bi-Retrograde is hallmarked with the Poinçon de Genève. Beating at 28,800 vph and offering 43 hours of power reserve, the 27-jeweled movement is equipped with a guilloché rotor. Retrograde day is provided at 6:00 and retrograde date is shown at 12:00. The three-piece case is polished and water resistant to 3atm. The sapphire caseback reveals the exquisite movement.

Minute Repeaters

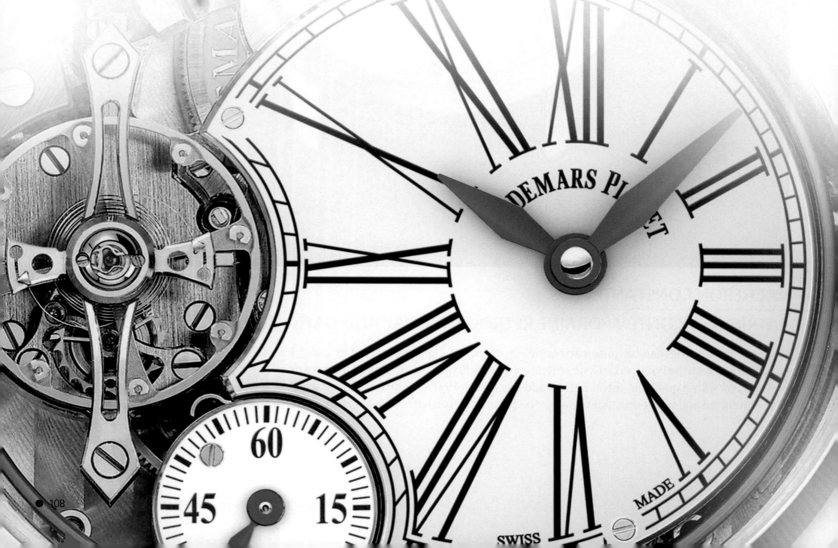

Audemars Piguet
Millenary Minute Repeater

Minute Repeaters

and Sonneries

The audible indication of time dates back to the very first mechanical clocks and is making a noteworthy comeback in the early 21st century. Over the past few years, we have been witnessing ever-increasing numbers of striking watches, especially the famous minute repeater models. Horologists are making the most of the latest advances in the fields of micromechanics, acoustics, information technology and materials science to offer timepieces chiming out ever more perfect tunes. While the tourbillon is becoming somewhat commonplace, minute repeaters and Grande Sonnerie (grand strike) models are asserting themselves as supreme emblems of exclusivity and rarity, while extremely wealthy collectors and connoisseurs rush to snap up these miniature technical marvels that express a blend of expertise and poetry.

STRIKING SYSTEMS WERE BORN AT THE SAME TIME AS MECHANICAL HOROLOGY.

From the emergence of the first monumental clocks, in around 1300, some were equipped with mechanisms serving to "sound out" the time at a distance and by night, whereas dials and hour hands did not appear until a century later. In the 14th century, the monumental clock with automatic chime, often equipped with jacks (automaton figures) that mimed the gesture of striking the hours, was part of the "adornment" of any self-respecting town. Wall clocks from 1350 onward and subsequently table clocks were also often equipped with chiming systems or striking bells, and sometimes enhanced by automatons. As soon as pocket-watches appeared, they were also often equipped with audible mechanisms striking the hours "in passing," meaning automatically, just like steeple clocks. The fashion for striking watches continued throughout the 17th century.

Moreover, a new type of striking mechanism was soon to make its appearance: the famous "repeater" watches serving to sound the time at will. The first was the "quarter repeater," striking the hours and quarter-hours. It was invented around 1675 by English watchmakers. The minute repeater first appeared between 1700 and 1710, notably in southern Germany. Striking on demand, it asserted itself as an extremely practical instrument for telling the time at night, as well as serving as a symbol of prestige, displaying its owner's power and wealth.

Repeater watches were much in favor during the 18th century. Various "programming" variations emerged, including ten-minute, half-quarter and five-minute repeaters. All these striking timepieces were equipped at the time with hammers striking a small bell. In the 1780s, the famous Abraham-Louis Breguet had the idea of replacing bells by gongs, meaning steel spring blades, initially placed across the back plate, and then wrapped around the movement inside the case. This crucial invention served both to achieve a clearer, purer sound and to reduce the thickness of the case.

From 1845 onwards, the invention of so-called "safety" matches, followed by the emergence of hands and dial with a luminescent coating, deprived repeating watches of their initial reason for being. Nonetheless, thanks to their technical sophistication and the magic of their sound, striking watches—and particularly minute repeaters—earned a definitive place in the horological hall of fame, right up among the rarest and most sought-after complications.

In the 20th century, by then miniaturized to wristwatch size—a change that involved a number of technical and acoustic challenges—watches with an audible indication of time asserted themselves as one of the greatest demonstrations of watchmaking know-how, and many observers indeed came to view them as the supreme complication.

The Minute Repeater

A minute repeater mechanism all by itself calls for over one hundred sometimes microscopic parts, which may be mainly grouped into the following categories:

HOUR, QUARTER-HOUR AND MINUTE PIECES

These three key components fulfill a two-fold role, both determining the angular position of the snails thanks to "feeler-spindles" positioned at one end, and setting off the right number of blows of the hammer thanks to the "rack" positioned at the other end.

THE STRIKING MECHANISM

The striking mechanism itself is composed of hammers striking on gongs.

GOVERNOR

To guarantee that the blows are struck at regular intervals, minute repeater watches are equipped with a governor (also known as a regulator) that controls the striking speed of the hammers.

ACTIVATION SYSTEM

This generally consists of a bolt or slide-piece that the user shifts along the caseband, both reloading the mainspring of the audible mechanism and triggering the striking mechanism.

SNAILS

The role of the snails is to reproduce below the dial the hour shown by the hands, in such a way that the striking mechanism can keep track at any time of the number of notes to be struck. Minute repeaters are equipped with three snails for the hours, quarter-hours and minutes, respectively.

Grande Sonnerie Watches

Making a minute repeater is a lengthy, meticulous and complex process that requires a combination of peerless technical skills and an excellent musical ear. It is reserved for an elite group of artisans with solid experience in this domain.

Grande Sonnerie (grand strike) models, are regarded as the most sophisticated of watches providing an audible indication of the time. They call for a number of parts far superior to the minute repeater (with which it is often associated). One of its main challenges lies in mastering energy. Whereas a minute repeater is reloaded each time the repeater slide-piece is activated, a Grande Sonnerie must at all times have sufficient energy to strike the desired number of notes (96 notes in 24 hours!) while consistently producing a sound of the same quality. To achieve this, it is generally equipped with two mainsprings: one for the timekeeping gear trains, and the other for the striking mechanism.

HAMMERS
The hammers are small steel levers pivoting on one axis and held in resting position by a spring. Once raised up by a "lifting lever", the hammer briefly strikes the corresponding gong in order to make it vibrate.

GONGS
Thin steel blades struck by the hammers. The gongs are wrapped inside the case all around the movement and fixed at one end. Depending on their length, they produce a lower or higher-pitched sound. The type of alloy plays a crucial role in terms of the quality of the sound.

"CATHEDRAL" GONGS
"Cathedral" gongs are gongs that are wrapped more than once—generally almost twice—around the movement and thereby generate a richer and deeper sound.

CARILLON CHIME
A carillon chime is a striking mechanism comprising at least three gongs.

WESTMINSTER CHIME
So-called Westminster chimes sound each quarter by successively striking four different gongs. Contrary to what their name might imply, they do not necessarily play the tune of the famous Big Ben—the clock on the British Parliament—originally composed by Handel.

REPEATING WATCHES
Watches equipped with a mechanism enabling them to sound the time on demand when a slide-piece or pushbutton is activated.

MINUTE REPEATERS
The most sophisticated type of repeater watch, the "minute repeater," sounds the time on demand to the nearest minute. These watches are generally equipped with two gongs: one low-pitched for the hour, and the other high-pitched for the minutes, with the quarters marked by a succession of high-low notes. Activating the minute repeater at 08:47 makes it sound 8 low notes, three double high-low notes and two high notes.

GRANDE AND PETITE SONNERIES (GRAND AND SMALL STRIKES)
Grande and Petite Sonnerie watches automatically sound the hours and quarters in passing, like a church steeple clock. Grande Sonnerie models repeat the hour before each quarter; while Petite Sonnerie models merely sound the hours or quarters (without repeating the hour at each quarter).

Minute Repeaters: Innovation at Work

In recent years, watchmakers seem determined to revive interest in minute repeater models by exploring new technical paths, with changes to construction or affixation of the chimes, alloys used for their fabrication, transmission of the sound, ways of activating the strike or the increasing water resistance of these ultra-sophisticated watches.

< Vacheron Constantin
Patrimony Contemporaine Ultra-Thin Caliber 1731

Presented as the world's thinnest minute repeater watch (8.9mm thick), the Patrimony Contemporaine Ultra-Thin Caliber 1731 by Vacheron Constantin is equipped with an in-house movement featuring an entirely silent striking governor.

> Breguet
Classique Grande Complication
Minute Repeater

Beneath its traditional appearance, the Classique Grande Complication Minute Repeater from Breguet conceals an entirely reinvented striking mechanism that uses new materials and innovative positions for the chimes, chime-holders and hammers.

Spero Lucem, *La Clémence*

The La Clémence watch from the new brand Spero Lucem is distinguished by an original and fascinating complication. When the minute repeater is striking, the hands "go crazy." At another strike of the hammer, they return to their appointed place and continue running normally again.

Bell & Ross
Vintage PW1 Minute Repeater

Bell & Ross has updated pocket-watches with an audible indication in this Vintage PW1 Minute Repeater model. The case is made of Argentium® 960, a silver alloy that is purer, brighter and more resistant to oxidation.

◀ **Christophe Claret**, *Adagio*

On this minute repeater model from Christophe Claret, Adagio, the cathedral chimes of the striking mechanism are equipped with a patented device that prevents the chimes from hitting each other while they vibrate under the hammers.

▶ **Blancpain**

Léman Minute Repeater Aqua Lung

Blancpain has opted to tackle one of the trickiest issues with minute repeater watches—the Achilles heel in their water resistance represented by the slide-piece—with its Léman Minute Repeater Aqua Lung, water resistant to 100m.

Bulgari, *Daniel Roth Carillon Tourbillon*

The Daniel Roth Carillon Tourbillon by Bulgari is distinguished by its striking mechanism with three hammers and three gongs, which are extremely rare on this type of model. It sounds the hours on a C note, the quarters with an E-D-C sequence, and the minutes on an E note.

Patek Philippe, *Ref. 5074*

In its Ref. 5074, Patek Philippe—which has a collection comprising around 13 minute repeater wristwatches—has combined this complication with a perpetual calendar. Its "cathedral" gongs feature a particularly lengthy resonance time.

Cartier, *Rotonde Minute Repeater*

On its Rotonde Minute Repeater watch, Cartier offers a new way of securing the gongs aimed at achieving perfect acoustic synergy between the case and the movement. The square-section gongs enable the hammer to regularly strike the same spot.

Jaeger-LeCoultre, *Reverso Répétition Minutes à Rideau*

Jaeger-LeCoultre's Reverso Répétition Minutes à Rideau invented a new sensation with a sliding shutter system that hides one of the faces of the case—while activating the strike mechanism.

Manufacture Royale, *Opéra*

The first creation from the hyper-exclusive brand Manufacture Royale, the Opéra tourbillon and minute repeater model stands out for its case, which resembles a metallic accordion bellows and acts as a resonance chamber for the gongs.

Minute Repeaters: Open-Hearted Mechanisims

On classical minute repeater watches, nothing signals the presence of this sophisticated complication—besides the presence of a small slide-piece or bolt on the side of the case. Following a general trend in haute horology, many watchmakers today prefer to reveal the mechanism by openworking the dial, or forgoing it altogether.

< **Christophe Claret**, *Soprano*

With its Soprano model, Christophe Claret associates a tourbillon with a minute repeater featuring a Westminster chime comprising four patented gongs and four hammers. This impressive feat is beautifully highlighted by stepped bridges inspired by the Charles X style.

∨ **Speake-Marin**
Renaissance Tourbillon Minute Repeater

Created by the independent master-watchmaker Peter Speake-Marin, the Renaissance Tourbillon Minute Repeater is fitted with an open-worked dial revealing the intricate inner workings of the mechanism, as well as the impressive tourbillon at 4:30.

de Grisogono
Occhio Ripetizione Minuti

Based on the principle of reflex cameras, the Occhio Ripetizione Minuti from de Grisogono features an aperture in place of a dial; when the minute repeater is activated, the twelve ceramic shutters open to reveal the movement and remain open while the cathedral chime is sounding.

Harry Winston
Midnight Minute Repeater

On the Midnight Minute Repeater from Harry Winston, with an off-centered hours/minutes display, a large "porthole" at 10 o'clock allows the wearer to admire the striking hammers, engraved with the initials H and W.

Seiko
Credor Spring Drive Minute Repeater

Signed Seiko, the Credor Spring Drive Minute Repeater features a skeletonized movement that is visible through an openworked dial and a sapphire crystal caseback. Its hammers are shaped from a special kind of steel forged by one of the most famous Japanese steelmakers.

Piaget
Emperador Coussin Automatic Minute Repeater

An embodiment of Piaget's virtuoso skills in the field of ultra-thin watchmaking, the Emperador Coussin Automatic Minute Repeater is equipped with a sapphire dial affording admirable views of the original architecture and highly refined finishing of the manufacture-made movement.

Badollet
Observatoire 1872 Minute Repeater

A wristwatch with a pocket-watch look, the Observatoire 1872 Minute Repeater from Badollet provides ample views of its hammers, gongs and inertial flywheel through an opaline openworked dial.

Audemars Piguet
Millenary Minute Repeater

Housed in an oval brushed titanium case, the Millenary Minute Repeater from Audemars Piguet stands out for its three-dimensional architecture that reveals the escapement, a double balance-spring and the hammers and gongs of the striking mechanism.

Vacheron Constantin, *Maîtres Cabinotiers Skeleton Minute Repeater*

Vacheron Constantin aims for maximum transparency—and slenderness—with its Maîtres Cabinotiers Skeleton Minute Repeater. The 3.3mm-thick movement comprises over 330 parts, most of them openworked, decorated and chased.

Bovet

Amadeo Fleurier 46 Minute Repeater Tourbillon Reversed Hand-Fitting

Equipped with a visible movement that is entirely engraved, the Amadeo Fleurier 46 Minute Repeater Tourbillon Reversed Hand-Fitting from Bovet can easily be transformed into a table watch, pocket-watch or pendant watch.

Repeater Watches: New Rhythms

The most popular repeater watches are currently the minute repeaters, which sound on demand the hours, quarter-hours and minutes. A few horologers have opted to reinterpret this classic model by revisiting the old-fashioned "five-minute repeater" (simpler to produce and hence less costly), proposing variations on the frequency and type of the strike, or creating repeaters that sound the time of another time zone.

> **Louis Vuitton**
>
> *Tambour Répétition Minutes*
>
> On the Tambour Répétition Minutes from Louis Vuitton, with a dual time zone, the minute repeater sounds the home time, not local time—a pleasant way to feel connected with close friends and family.

> Edox
Cape Horn 5 Minute Repeater

Equipped with an additional Dubois Dépraz model, the Cape Horn 5 Minute Repeater by Edox is distinguished by its openworked and engraved movement visible from both sides of the case—and by its price that is two or three times lower than that of other models of this type.

∧ Claude Meylan
Répétition 5 (Ref. 8731)

The Répétition 5 (Ref. 8731) by Claude Meylan features a movement entirely openworked and engraved by hand. Pressing the pushbutton at 8 o'clock makes the watch sound the number of hours, followed by the number of five-minute intervals elapsed since the previous hour.

< Kari Voutilainen
Masterpiece N° 7

On his one-of-a-kind model, the Finnish-born watch-maker offers a minute repeater striking not only hours/quarter-hours/minutes, but also hours/ten minutes/minutes, a system designed to enable a more intuitive division of time.

> deLaCour, *Birépétition minutes*

Equipped with a Christophe Claret movement, the Birépétition minutes from deLaCour strikes the hours, quarter-hours and minutes for either the local time or a second time zone.

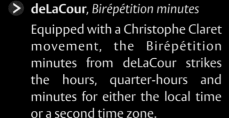

∧ Seiko
Credor Spring Drive Sonnerie

The Japanese brand Seiko has revived the antique hour repeater mechanism in a high-end model named Credo Spring Drive Sonnerie. The mechanism strikes in three modes: the number of hours on each hour; three strikes every three hours (3, 6, 9 and 12 o'clock); and silent mode.

Grand and Small Strikes

Like a church bell on the wrist, grand and small strike systems, also called grande et petite sonnerie, mark the hours and quarter-hours in passing (with grand strikes repeating the hours each time). This extremely rare function is the prerogative of ultra-complicated wristwatches (see chapter on Multi-complications). Having first appeared in wristwatch form in a model presented by Philippe Dufour in 1992, it has also cropped up since then in some highly sophisticated watches where it most often joins the minute repeater—sometimes in chiming the famous melody of Big Ben.

F. P. Journe, *Sonnerie Souveraine*

Possessing grand strike and minute repeater capabilities, the Sonnerie Souveraine is distinguished by its extreme functionality and security assured by ten patented devices. F.P. Journe actually designed it "so an eight-year-old child could use it without damaging it"—which required a high degree of technical sophistication in the conception and design.

A. Lange & Söhne, *Zeitwerk Striking Time*

The Zeitwerk Striking Time from A. Lange & Söhne strikes each quarter-hour in passing with a high note, then the hour with a low one—thus reproducing in slow motion the famous tune from the beginning of Beethoven's Fifth Symphony.

Philippe Dufour
Grande Sonnerie

Launched in 1992, the Grande Sonnerie from Philippe Dufour—one of the preeminent members of the Académie Horlogère des Créateurs Indépendants—was the first minute repeater watch with grand and small strike.

Chopard, *L.U.C Strike One*

On its L.U.C Strike One, Chopard presents a mechanism striking each hour with a single note (hence its name), while an opening in the dial serves to reveal a small hammer at work.

Bulgari, *Grande Sonnerie Perpetual Calendar*

The Bulgari Grande Sonnerie Perpetual Calendar has a chime playing the Westminister tune on four notes. It is also equipped with two barrels, of which one is exclusively dedicated to the striking mechanism and guarantees a 24-hour power reserve in grande sonnerie mode.

Audemars Piguet, *Grande Sonnerie Carillon*

Presented at the end of the 1990s, the Grande Sonnerie Carillon by Audemars Piguet is distinguished by its dynamograph system indicating the real torque delivered by the barrel. It has a power reserve of 20 hours in Grande Sonnerie mode and 60 hours in Petite Sonnerie mode.

Watches with Automatons

In terms of a grand visual and acoustic spectacle, watches with automatons, which are extremely rare, unquestionably occupy an echelon all to themselves.

> **Jaquet Droz**, *The Bird Repeater*

Heir to the famous 18th-century Jaquet Droz automatons, The Bird Repeater wristwatch features a pair of birds on a nest in front of a stream—a three-dimensional tableau that springs to life each time the cathedral-gong minute repeater is activated.

v **Bulgari**, *Commedia dell'Arte*

The Commedia dell'Arte by Bulgari sets the stage for the three main protagonists in this traditional form of Italian theatre. When the "cathedral gong" minute repeater is activated, the central figure (whose arm shows the minutes) raises his right arm, while the other figures begin a lively dance.

^ **Van Cleef & Arpels**, *Lady Arpels Poetic Wish*

Signed Van Cleef & Arpels, the Lady Arpels Poetic Wish combines a five-minute repeater with three automatons—a young girl, a tiny mother-of-pearl cloud and a kite—moving on request to sound the hours and minutes in music.

> **Ulysse Nardin**
> *Carnival of Venice Minute Repeater*

On the Carnival of Venice Minute Repeater by Ulysse Nardin, two small figures standing on the Rialto Bridge raise their masks each time the repeater chimes the hours, quarters and minutes. An 18-piece limited edition.

< **Bulgari**, *Daniel Roth collection, Il Giocatore Veneziano*

Inspired by a painting by Caravaggio, Il Giocatore Veneziano from Bulgari's Daniel Roth collection associates a minute repeater mechanism (with cathedral chime) with an automaton that throws two dice with a total of 504 possibilities. The automaton movement may be activated either at the same time as the striking mechanism, or on demand.

> **Bovet**
> *Amadeo Fleurier 44 Minute Repeater Tourbillon Triple Time Zone Automaton*

The Amadeo Fleurier 44 Minute Repeater Tourbillon Triple Time Zone Automaton watch by Bovet features a dial featuring an automaton with bells reminiscent of the first jaquemart (jack) mechanisms.

Feminine Melodies

In a still tentative trend, we have recently seen the first watches with minute repeater or "in passing" strike mechanisms created especially for women.

Breguet
Reine de Naples Sonnerie au passage 8978

To celebrate the 200th anniversary of the wristwatch created for Caroline Murat, Queen of Naples, Breguet presented the Reine de Naples Sonnerie au Passage 8978, with a mechanism that marks each full hour with two strikes repeated three times.

Ellicott, *Lady Tuxedo Midnight*

Ellicott's Lady Tuxedo Midnight model is equipped with a self-winding movement striking the hours and quarter-hours on request. Its sensual case, partially or entirely set with diamonds, is crafted in rose gold, white gold or steel.

Patek Philippe, *Ladies First Minute Repeater Ref. 7000*

The first Patek Philippe minute repeater wristwatch exclusively dedicated to women, the Ladies First Minute Repeater Ref. 7000 is distinguished by its ultra-thin self-winding movement and its pure aesthetic, featuring a rose-gold case set with diamonds.

Musical Watches

To ring out the melody of time, several watches incorporate miniature mechanisms inspired by historical music boxes. They feature a varied repertoire ranging from great classic tunes to modern hit songs.

⌄ Ulysse Nardin, *Stranger*

On its mechanism, clearly visible through the dial side, the Stranger watch by Ulysse Nardin plays the tune of "Strangers in the Night" every hour or whenever the user presses a button. The crown with its integrated pusher serves to adjust various functions, including winding, moving the date forward or back, as well as setting the time.

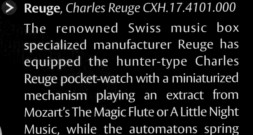

› Reuge, *Charles Reuge CXH.17.4101.000*

The renowned Swiss music box specialized manufacturer Reuge has equipped the hunter-type Charles Reuge pocket-watch with a miniaturized mechanism playing an extract from Mozart's The Magic Flute or A Little Night Music, while the automatons spring to life around the fountain.

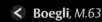

‹ Boegli, *M.63*

Boegli has equipped its M.63 Far West watch with a "mechanical music movement" playing the famous tune "For a Few Dollars More" by Ennio Morricone. The brand also offers various wristwatches and pocket watches paying tribute to the great composers such as Verdi, Mozart and Vivaldi—even "God Save the Queen."

A. LANGE & SÖHNE

LANGE ZEITWERK STRIKING TIME – REF. 145.029

This member of the Lange Zeitwerk family is endowed with a chiming mechanism that is visible on its dial. The Lange Zeitwerk Striking Time is the first Lange wristwatch with an acoustic signature. It strikes the quarter-hours with high-pitched tones and the full hours at a lower pitch. The chiming mechanism consists of two black polished steel hammers that are integrated in the dial layout on either side of the subsidiary seconds. The manually wound L043.2 caliber is crafted to the most exacting Lange quality standards and largely decorated and assembled by hand. The sonorous timepiece comes in a 44.2mm white-gold case with a black dial or in a limited edition of 100 pieces with platinum cases and rhodium-plated dials.

A. LANGE & SÖHNE

LANGE ZEITWERK STRIKING TIME – REF. 145.032

Combining a sonorous minute repeater with the contemporary visual appeal of two jumping-numeral apertures, the pink-gold Lange Zeitwerk Striking Time creates a multi-dimensional, multi-sensory spectacle of unobstructed clarity. In fact, the 44.2mm timepiece goes one step further, revealing the L043.2 caliber's polished steel hammers on either side of the 6:00 subsidiary seconds. The hammers, which indicate the quarter-hours and hours via high- or low-pitched tones respectively, may be silenced with ease by a push of a button at 4:00, preventing any disturbance if the wearer so desires. Apertures at 3:00 and 9:00 display the hours and minutes by way of three numeral discs that advance precisely with each passing minute thanks to the movement's patented constant-force escapement.

AUDEMARS PIGUET

JULES AUDEMARS OPEN-WORKED MINUTE REPEATER – REF. 26356PT.OO.D028CR.01

The Jules Audemars Open-worked Minute Repeater with jumping hour and small seconds is powered by Audemars Piguet's in-house, open-worked, hand-wound Manufacture caliber 2907, which is comprised of 420 parts and beats at 21,600 vph. This timepiece has a 950 platinum case and folding clasp, with a sapphire crystal exhibition caseback. It features an open-worked white dial and is secured to the wrist with a blue alligator leather strap. It displays jumping hour through dial aperture, minutes and small seconds, and there is a minimum guaranteed power reserve of 72 hours.

AUDEMARS PIGUET

MILLENARY MINUTE REPEATER – REF. 26371OR.OO.D803CR

The new Millenary Minute Repeater in rose gold is an exceptional, hand-wound wristwatch with an oval case and three-dimensional architecture that complements a refined and sophisticated movement highlighted by Audemars Piguet's own escapement, double balance spring, striking mechanism and gongs. The result is a visual and aural delight. The distinctively shaped case provides an instantly recognizable setting for the gold and enamel subdials, which invite closer inspection of the remarkable mechanism within. The manual-wound caliber is distinguished by the atypical construction of the regulating organ, which is composed of not just one balance spring, but of two placed top to tail.

BELL & ROSS

PW MINUTE REPEATER SKELETON DIAL

The skeletonized dial on the PW1 Minute repeater allows the wearer to see the inner workings of the Dubois Dépraz hand-wound movement. Complete with a power reserve of 56 hours, this five-minute repeater is unique in that it is activated upon request by pressing the button on the left-hand side of the timepiece. A low-pitched sound radiates every hour followed by a series of doubled low-pitched and high-pitched tones for every five minutes that pass. This musical masterpiece is finished with a shining argentium case and is also available with a ruthenium finish on the dial.

BLANCPAIN

CARROUSEL MINUTE REPEATER FLYBACK – REF. 2358-3631-55B

This unique model is truly a work of art. The masters at Blancpain include their well-known one-minute flying carrousel at 6:00. The minute repeater features a cathedral gong with blades that wrap one and a half times around the movement in order to emit a sound of surprising quality. The third function, which also stems from Blancpain's classic legacy, is a chronograph, whose pushbutton at 4:00 enables the wearer to restart a time measurement while a first measurement is in progress. This unique arrangement, featuring a 30-minute counter at the center of the caliber, demonstrates the horological mastery of the Blancpain movement design engineers.

CHRISTOPHE CLARET

ADAGIO – REF. MTR.SLB88.024-824

Adagio is an astounding minute repeater with cathedral gongs, featuring a large date at 6:00 and a second time zone between 1:00 and 2:00. As a tribute to his manufacture's first caliber introduced over 20 years ago, Christophe Claret has designed Adagio as a minute repeater—chiming the hours, the quarter-hours and minutes after the quarter-hours on demand—thus enabling the time to be heard rather than read. The in-house Caliber SLB88 is composed of 455 components. The cathedral gongs of the striking mechanism feature a patented device that prevents them from touching when they vibrate under the hammer strikes.

CHRISTOPHE CLARET

SOPRANO – REF. MTR.TRD98.020-028

The Soprano watch associates two of the finest horological complications: a 60-second tourbillon and a minute repeater with Westminster chime, four patented cathedral gongs and four hammers. The minute repeater mechanism is highlighted by stepped bridges inspired by the Charles X style, a favorite of Christophe Claret. This exceptional caliber is housed inside a round case that subtly marries precious metals and titanium. Tradition and modernity are united in this creation, entirely in keeping with the demands of ultra-high-end watch making developed by the Manufacture Claret.

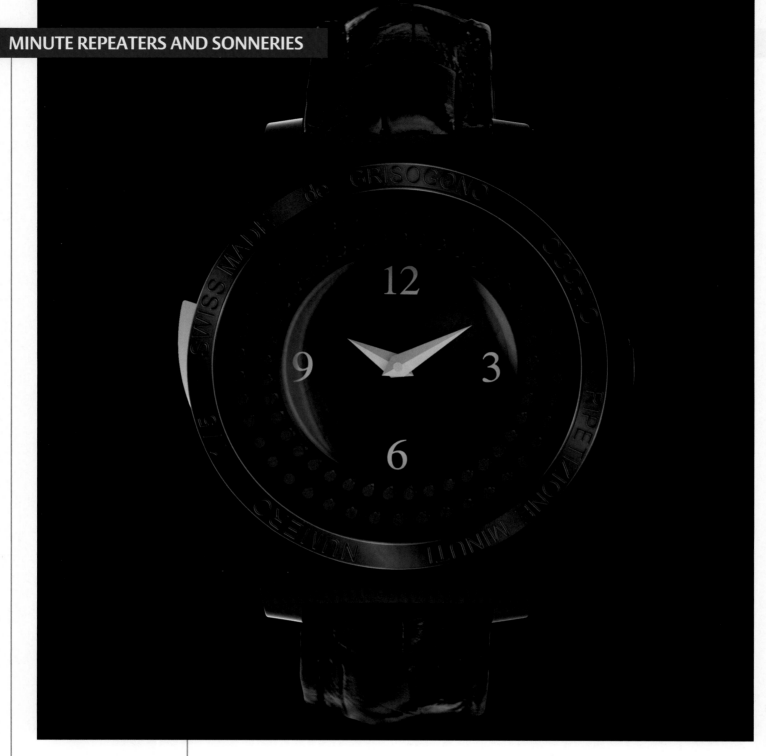

de GRISOGONO

OCCHIO RIPETIZIONE MINUTI – REF. OCCHIO S03

Housed in a blackened titanium case accented by 88 black diamonds on the bezel, this self-winding timepiece combines the classicism of a minute repeater with the contemporary ingenuity of an innovative shutter system. Upon the revelation of the movement via the slide-activated opening of the dial's 12-blade diaphragm, this monochrome wristwatch serenades its wearer with a three-chime cathedral gong that marks the hours, quarter-hours and minutes.

F.P. JOURNE

SONNERIE SOUVERAINE

Operating a chiming watch has always been risky, as the slightest mistake in operation can damage the movement. To meet the demand for an easy-to-use timepiece, construction of the Sonnerie Souveraine required ten patents on new mechanisms, including the barrel, power reserve indicator, winding and setting systems and chiming gongs. In this grand strike wristwatch, a single mainspring provides enough energy for 24 hours of grand strike (96 full chimes in passing). The chiming functions alone use up almost 60% of the mainspring's energy, and in silent mode without the chime, the movement runs for five days. Constructing this movement has been a permanent quest to maximize mechanical efficiency. The result is a low-tension movement with mechanisms that have to be very finely adjusted to ensure unfailing chimes 35,040 times a year.

GUY ELLIA

REPETITION MINUTE ZEPHYR

The ambassador of Guy Ellia's complicated timekeeper, the Répétition Minute Zephyr with its complications and intricate manufacturing process, is considered a particularly exceptional piece. The Caliber GEC 88 was created by Swiss manufacture Christophe Claret and features a power reserve indicator, minute repeater function and five time zones with day/night indication. The movement beats at 18,000 vph in a solid, convex, high-resonance sapphire crystal case of impressive proportions (53.6x43.7x14.8mm). Mounted on an alligator strap with a folding buckle, this model is also available in pink gold or titanium.

JAEGER-LECOULTRE

MASTER GRANDE TRADITION MINUTE REPEATER – REF. 5012550

The result of many years of research by Jaeger-LeCoultre, this watch incorporates the latest progress in the area of sound. Faithful to the spirit of the new Grande Tradition collection, its mechanical caliber in nickel-silver, with a power reserve of two weeks, is entirely decorated by hand. Once again, this model reflects the Grande Maison's watch making artistry and mastery. The exclusive minute repeater is only available in a limited series of 100 timepieces in 18K rose gold.

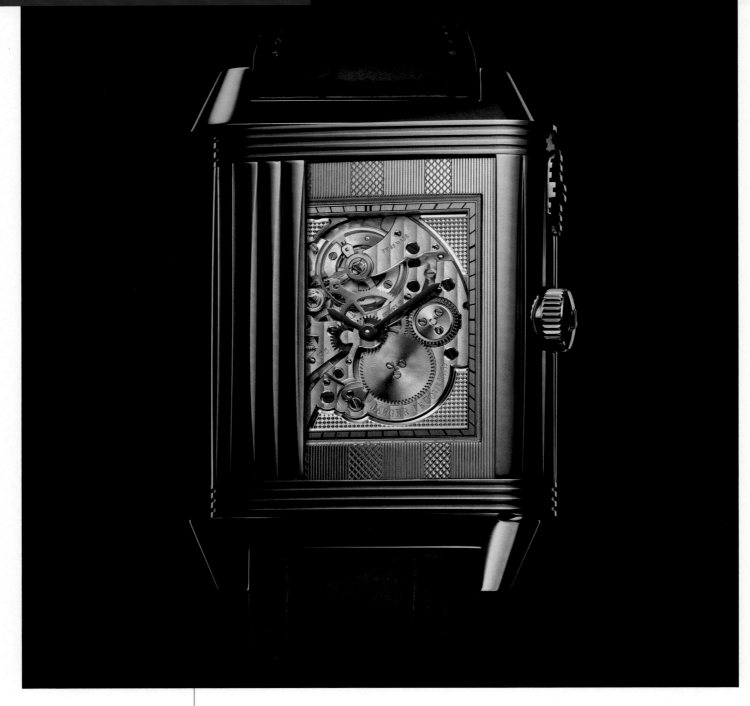

JAEGER-LECOULTRE

REVERSO REPETITION MINUTES A RIDEAU – REF. 2353520

This haute horlogerie watch incorporates a curtain that slides to active a minute repeater function. This exceptional timepiece enables the Grande Maison to provide yet another demonstration of its expertise in the field of sound. The masters at Jaeger-LeCoultre have integrated trebuchet hammers, monobloc gongs made from a secret-formula alloy, and a silent regulator to achieve this model. The sound produced by the Reverso Répétition Minutes à Rideau is the result of long research carried out in the Manufacture's laboratories. Housed in the iconic Reverso case, this piece reflects the perfect blend of technical innovation and fine aesthetics.

PARMIGIANI

TORIC CAPITOLE – REF. PFH276-1203300-HA1441

Immediately standing out with its ingenious sectorial display of the hours and minutes, this 18K white-gold timepiece, with polished finish and fully hand-engraved and -knurled bezel, serenades the exquisite sheen of its white mother-of-pearl dial with a two-gong "serpents" cathedral chime sounding the hours, quarter-hours, and minutes. Worn on an Hermès alligator leather strap, the watch is driven by the 397-component hand-wound PF 321 caliber with 45-hour power reserve. The demonstrative timepiece is finished with a sapphire crystal caseback engraved with the model's individual number.

Multi-Complications

Jaeger-Le Coultre
Hybris Mechanica à Grande Sonnerie

Christophe Claret

Founder of Christophe Claret

"When I have dreams, I make them come true"

Soleil d'Or, Christophe Claret's manufacture in Le Locle

Watchmaker-entrepreneur Christophe Claret has been **CREATING WATCHES FOR PRESTIGIOUS BRANDS FOR MORE THAN 20 YEARS.** In 2009, he founded his own brand, under his own name. Adventurousness, independence and a passion for playful horology guide his very exclusive production. Increasing steadily, the brand's production will reach 200 watches this year. In 2013, Christophe Claret launched the emblematic Kantharos model, which characteristically features an unusual juxtaposition of complications: in this case a chronograph with a striking mechanism. In 2014, Christophe Claret completes its trilogy of watches devoted to gaming by introducing the Poker watch. To top it off, the number of retail outlets that offer Christophe Claret watches is expected to double in the next three years.

You produce watches for the biggest Swiss watchmaking brands. Why did you launch your own brand?

I wanted to go on that adventure. How stimulating, to follow my own path and create my own products! At the same time, I continue to work for third parties.

Does having your own brand provide the opportunity to undertake projects that would be difficult to realize for others?

Yes, true innovation is very frightening. It entails technical and economic risks. There is no guarantee that a pioneer will succeed, since it involves tracing the path that the others will follow. For example, the technique may be a failure, and lose money. Those who come later will already know about the potential pitfalls and use simpler solutions. In difficult economic times, brands, particularly publicly traded ones, are even more reluctant to take risks. With my own brand, I create watches that would be considered too daring for others.

What obstacles have come up since Christophe Claret's launch?

Actually, I expected it to be a lot harder! Over 20 years, the brands that I supplied told me how hard it was: how late their clients paid, or about the difficulties of distribution. They were actually trying to discourage me. Since then, I've realized that a brand takes many fewer risks than a manufacture. The latter conceptualizes, realizes and perfects an innovative product. If it fails, the manufacture loses a large investment in research and development. The profit margins are also smaller, and tighter. Whereas a brand buys a finished product, and can easily change suppliers if it doesn't work out. Today, I am familiar with and able to handle both roles: the manufacture and the brand.

Christophe Claret,
Founder of
Christophe Claret

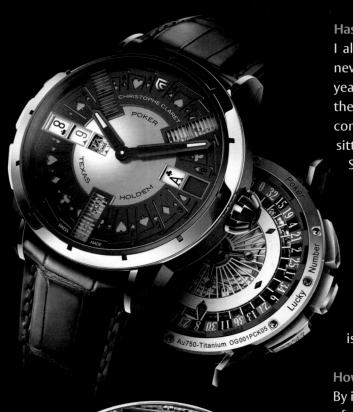

Has Christophe Claret's independence influenced its success?

I always swore that the day I launched my own brand, I would never depend on anyone else for cases, dials or hands. For 20 years, I had to deal with the fact that if there were delays with their other suppliers, our clients would not be able to put our very complicated movements in their cases. Our calibers were thus left sitting on tables, waiting for the case, for example, or the hands.

So I chose vertical integration. In other words, we produce all of our components, such as the cases, dials and hands. For the raw materials, such as ceramic or titanium, we plan ahead as much as we can, to avoid any slowdown in production.

Our manufacture encompasses about 30 different professions. This almost-total autonomy comes at a price. On the other hand, I can rely on our own resources, so I can guarantee that we can honor our clients' orders. I do not lose a single sale due to lack of hands, dials or cases. My credibility is at stake, so it is well worth the required investment.

How does the Christophe Claret brand distinguish itself?

By its playful, interactive watches. I have nothing to prove in terms of complexity, and I consider horology to be like a game. I want businessmen to relax and amuse themselves a bit with our watches! We also offer more technical, traditional watches such as the Adagio, Soprano or Kantharos. They pay homage to watchmaking and artisanal professions. At the same time, we create extremely technical watches, such as the DualTow or the X-TREM-1. With these models, I am charting unexplored territory. For example, I inserted two magnets into the movement of the X-TREM-1, even though magnetism is the watchmaker's sworn enemy. I love this kind of challenge.

⌃ Poker and movement

❯ 21 Blackjack

What are your criteria for developing watches?

Technique is first and foremost, for reasons of reliability. Then we integrate a coherent aesthetic adapted to our needs. Our complications are always really useful, as well as playful. We are developing a new vision of horology. For example, in my opinion, placing two tourbillon movements right next to each other shows a certain lack of technical flair. However, that changes if they are connected via a differential gear that calculates their average functioning. The precision that one gains in that scenario makes the complication useful.

How will you enhance your collections this year?

Along the lines of the playful watches, the Poker follows the 21 Blackjack and the Baccara. It combines aesthetic technique with highly advanced technical features, at a more moderate price. At BaselWorld, we are also presenting a true feminine complication, a complication that does not exist on men's models! And we are introducing a men's watch with a traditional detent escapement. . It took seven years to develop and is protected by several patents.

What function do you dream of creating?

When I have dreams, I make them come true! Sometimes they help me as well. For example, in the gaming trilogy, we developed the 21 Blackjack and the Baccara first because the Poker required a more complex solution. All the cards are used in a poker game, in numerous combinations. Seven of us searched for the solution. It came to me in a dream three years ago. That's where I get my inspiration, at night, or when I'm driving my tractor in France. More than from watchmaking itself, my ideas generally stem from nature, mechanics or aerospace science.

Which Christophe Claret watch best represents the brand?

The Kantharos. It combines a striking mechanism and constant force with a highly precise chronograph. Although highly technical, it remains interactive and playful because of its unusual combination of a chronograph and an audible indication. An original choice, yet a very logical one! Indeed, a strike or chime—or a whistle, or a cannon—announces the beginning of any sporting match. With its 670 components, this watch brings together the chronograph and the striking mechanism. A number of highly innovative technical challenges made this achievement possible.

Your watches are very complex, intended for a certain type of client. How large is your estimated customer base?

We already have faithful customers in various countries. Today, fortunes are being made in emerging countries, such as China, Latin America or India. These collectors and watch lovers could potentially number in the tens of thousands. Additionally, there are many more brands than there were 25 years ago that produce complicated watches in order to satisfy and take advantage of the growing evolution of this new clientele.

How will you develop your distribution to reach out to these potential clients?

Today, the Christophe Claret brand is sold in 14 retail locations. We expect to double that in the next three years. For example, we are looking for a partner in mainland China. We are also developing our brand in the Middle East. The models we offer are adapted to this market, with watches such as the Soprano, Kantharos or Rialto.

Kantharos

Ricardo Guadalupe

CEO of Hublot

"We produce watches that are different, connected with the future"

HUBLOT CULTIVATES "THE ART OF FUSION": of materials, of innovation and tradition, and even of marketing and sponsoring. In harmony with this concept, CEO Ricardo Guadalupe announces the highlights of 2014: Hublot's role as official timekeeper of the FIFA World Cup, the brand's new ambassador, Brazilian soccer legend Pelé, the continued enlargement of the manufacture, 100 stores by the end of the year and new Ferrari limited editions.

⌃ Ricardo Guadalupe and Pelé, Hublot ambassador and Brazilian soccer legend, in front of the Hublot store in Rio

What makes Hublot unique?

Without a doubt, the concept of the art of fusion within horology. This operates on several levels: product development, innovation and even marketing. We want to develop and create horological products that are different, with new technologies and materials, connected with the future but grounded in respect for horological tradition. In 1980, Hublot was the first brand to elevate the rubber strap by pairing it with gold. Since then, it's become a watchmaking standard. We also introduced ceramic to high-end watchmaking, and it is now a reference material, like steel or gold.

This mission to stand out extends to our movements as well. For example, the UNICO movement reinterprets the chronograph complication. Its mechanism appears on the dial side of the watch, which also makes it different.

How does the mixture of materials reflect the art of fusion?

Hublot carries out a great deal of research on materials. The result of this research, our 18-karat Magic Gold, has a hardness of 1,000 Vickers. This significant innovation met with great commercial success. Producing and working with Magic Gold demanded further development and investments in machine tools. Laser, ultrasound and high-pressure ovens—these high-tech instruments didn't exist anywhere. We will be integrating them [into our process]. So the manufacture will expand: we are constructing a second building next door to the one we have now. We expect to finish it in summer 2015. A bit larger than the first, it will house the production of cases and components—notably in very technological materials—and movements. Once the components are machined and pre-assembled, the final assembly will take place in the existing manufacture. Hublot will also need a larger work force.

Ricardo Guadalupe, CEO of Hublot

Jean-Claude Biver, LVMH Group, President of Watch Division and Chairman of Hublot

Cathedral Minute Repeater Tourbillon and column-wheel chronograph

What themes have guided Hublot's evolution over the last few years?

For about seven years, the brand has been positioning itself in terms of marketing and sales. For three years, Hublot has been integrating new watchmaking specialties. Behind a strong marketing plan, the brand is demonstrating its horological substance, as a true manufacture. About thirty professional specialties have joined Hublot: in research and development and innovation, but also in production, profile turning, machining and surface treatments. Hublot already produces its own chronograph movement entirely in-house, as few other brands do. We also produce all of our complications in-house, including, last year, about 400 tourbillons—totally developed, produced and assembled here in our manufacture. In 2013, manufacture-made movements equipped 30% of our 38,000 watches. In five years, our goal is to produce 80% of our movements in-house, for a total volume of around 40,000 watches. That represents a huge challenge. The industrialization and integration phase for these professions will take a natural course over several years.

What is your favorite complication at Hublot?

The Cathedral Minute Repeater Tourbillon and column-wheel chronograph, launched in 2012. Highly exclusive, with a production of just 20 watches per year, it demonstrates our expertise with its exceptional sound volume, obtained using carbon fiber.

The last few years, numerous partnerships have raised the brand's profile. Do they reflect a different, more costly strategy than that of other brands?

All the brands in our sector devote the same percentage of their budget to marketing. We have chosen to concentrate on sponsorships, allocating 25% of our marketing budget. Unlike the majority of watchmaking brands, Hublot sponsors not just a single sport, but several fields. Our clients, an active group, can appreciate Formula 1, soccer matches, the music of Depeche Mode and skiing the slopes of Courchevel or Verbier. Our marketing is as diverse as the tastes of our clientele, because Hublot forms an integral part of their world. We go where our clients are.

Ferrari represents an important partnership. Are the results as productive as you had hoped?

We accompany Ferrari at numerous events, from the smallest to the largest, a total of almost 160 per year. Besides the increased presence for us, the success of this partnership can be measured by the two or three Ferrari limited editions we produce each year. Highly exclusive, they are often ordered in advance.

In June 2014, Hublot will be the official timekeeper for the FIFA World Cup. How will you make the most of this partnership?

The World Cup offers us a global presence. At BaselWorld 2014, Hublot unveiled a special edition of the FIFA World Cup Big Bang. During the match, the fourth official's board will display the name of the brand. This board indicates any changes in the players and any additional time in the match. This guarantees us airtime of 20 to 30 seconds per match, for a total television audience that numbers in the billions. We have also opened a boutique in Rio, and we are creating the Hublot Palace in a hotel in Copacabana. It will welcome end customers, retailers, partners, journalists and stars such as Pelé, our new ambassador as of January 2014. Hublot will also use the Hublot Palace to promote Swiss wines, a favorite endeavor of our Chairman Jean-Claude Biver. We will offer tastings of wines from Vaud, Switzerland, and a three-star chef will prepare characteristic Swiss food during the pre-World Cup dinner. All of these event-related activities will, naturally, increase our profile, particularly among young people, our future clients.

What was the highlight for Hublot in 2013?

Hublot had a record year, as it does almost every year! In terms of units sold and sales figures, the brand is growing quickly and finds itself among the front-runners. This has happened without any compromise in quality. We have also opened 14 boutiques in 2013, more than one a month! That brings the number of Hublot stores to 70 as of the end of 2013. In just four years, we have opened points of sale in the key global markets.

What are your priorities in terms of markets?

Hublot has had great success in Western Europe, the Americas, Eastern countries and Japan, with an incredible boom in business. Over the next five years, we are concentrating our efforts on China: we want to increase our market share there. To give you some background, exports to China represent close to 30% of Swiss watchmaking sales. At the present, China represents just 6% of our sales, so we have a lot of potential. Though their parents may have preferred classic, three-handed watches, the younger generations in China appreciate our brand's strong personality.

Besides the special FIFA watch, what are you proposing for 2014?

We are adding to the Classic Fusion line, which is our second mainstay along with the Big Bang collection. We are combining complications, such as a tourbillon chronograph and a minute repeater tourbillon. We have two watches made 100% in-house, which allows us to offer them at an accessible price, in a sport-chic aesthetic. We are also coming out with new models for women.

⌃ Hublot store in Rio

⌄ Ferrari Limited Edition 2014

Multi-Complications

With movements comprising several hundred parts and accommodating a vast array of functions within a minimum amount of space, multi-complication watches constitute the quintessence of the watchmaking art, the supreme demonstration of expertise, the signature of exceptional brands. The progress in the domain of computer technologies—for both design and production—have led to significant

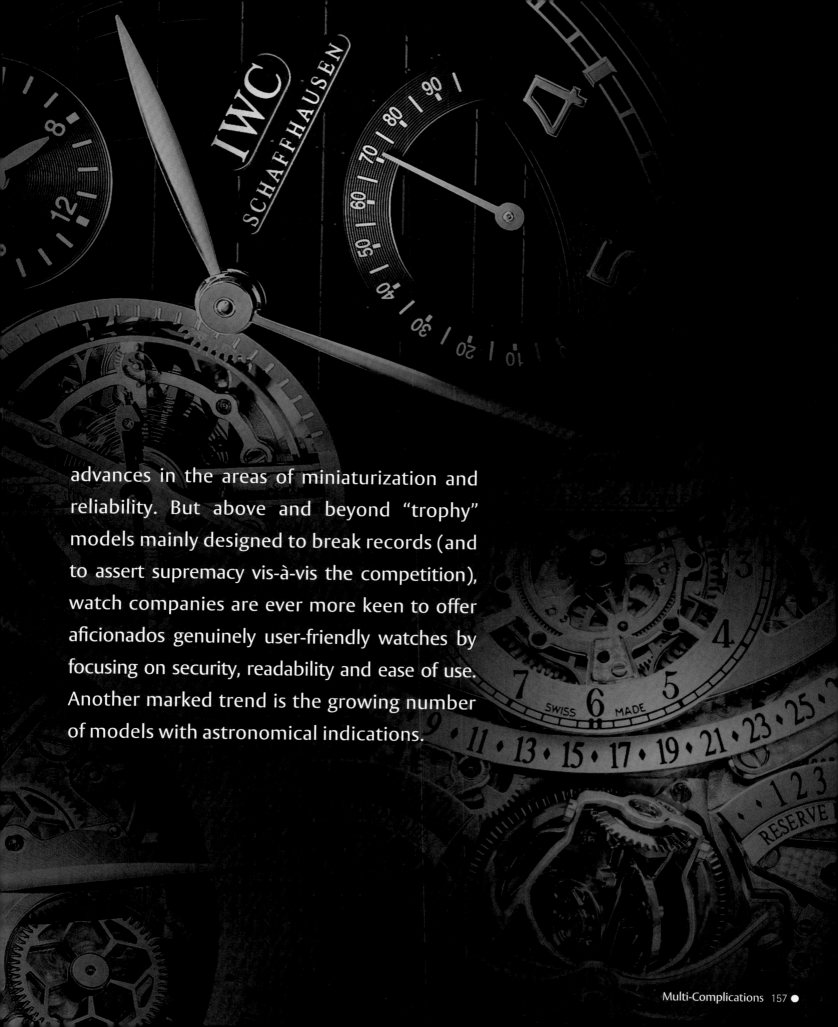

advances in the areas of miniaturization and reliability. But above and beyond "trophy" models mainly designed to break records (and to assert supremacy vis-à-vis the competition), watch companies are ever more keen to offer aficionados genuinely user-friendly watches by focusing on security, readability and ease of use. Another marked trend is the growing number of models with astronomical indications.

HOROLOGY DID NOT EVOLVE IN A PURELY LINEAR FASHION FROM SIMPLEST TO MOST COMPLICATED TIMEPIECES.

The minute display, which today seems one of the most basic functions, only became widespread at the end of the 17th century, with the invention of the regulating balance spring. However, some pocket watches already featured highly complex mechanisms by this point. As for the seconds display, it would not come along for another century. Moreover, the history of horology is marked by several "setbacks," due to the shrinking size of the mechanisms. This is why the first portable clocks and pocket watches of the Renaissance were more rudimentary, technically, than the large astronomical clocks of the Middle Ages—and even the most sophisticated wristwatch today would not be able to display as many functions as an ultra-complicated pocket-watch.

The first explicit mentions of mechanical clocks appear at the beginning of the 14th century. Though we can reasonably assume that artisans began with simple mechanisms, the only clocks we know from that period are quite complex, such as the clock tower with the astronomical dial that Roger Stoke constructed for the Norwich cathedral between 1321 and 1325, and of course the "star" of the era, a marvel of its time, Giovanni de Dondi's astrarium in Padua, completed in 1364.

During the late Middle Ages and the Renaissance, the Church and political authorities encouraged and financed monumental clocks that boasted a large number of astronomical and astrological indications, equipped with audible strike systems that were often quite spectacular. Among the most famous examples are the astronomical clocks of the Lund cathedral, in Sweden (1380), the Saint-Etienne cathedral in Bourges, France (1424) and the Saint Mary of Rostock church in Germany (1472).

The first pocket-watches, at the end of the 15th century, possessed only an hour hand. In the 16th century, artisans developed astronomical and astrological displays, then added more and more functions—alarm, calendar, strike in passing and on demand—as they competed to create watches that contained the largest number of complications.

The most famous example of this phenomenon is the Breguet N° 160, better known as the "Marie-Antoinette," whose creation and history are shrouded in history. In 1783, Abraham-Louis Breguet, already a favorite of Louis XVI and his wife, received a commission from an officer of the queen's guards: to create a watch that would integrate all the complications and refinements of the time. Marie Antoinette, guillotined in 1793, would never see this beauty, as it was completed in 1827, under the direction of Breguet's son. This automatic-winding watch (referred to as a perpétuelle) contained complications such as a minute repeater, complete perpetual calendar, equation of time, power reserve indicator, metallic thermometer, independent large seconds and small seconds hand—not to mention the many technical feats such as a double pare-chute, a gold balance spring and several components in sapphire crystal, all within its gold case and rock crystal dial. The Marie-Antoinette passed through several hands, and was stolen from a Jerusalem museum in 1983. The Breguet manufacture created an identical reproduction of the timepiece, but in the autumn of 2007, the original suddenly reappeared in the museum's collections under mysterious circumstances.

In 1925, the New York banker Henry Graves, Jr. commissioned Patek Philippe to create the most complicated watch ever for his personal collection. Comprised of over 900 components, this double-sided pocket watch includes 24 complications, including a grand strike on four gongs and alarm on a fifth gong. The title of "the world's most complicated watch" was long disputed between the Graves and another horological masterpiece, the Leroy 01 (1904).

Its double-sided case contains 20 complications related to time measurement—equation of time, chronograph, the sky above the Northern and Southern Hemispheres, grand strike, etc.—and five non-horological functions: hygrometer, thermometer, barometer, altimeter and compass. This inclusion has given rise to a dispute between those who grant it 25 complications and those—partisans for the Graves—who contest that it has "only" 20.

In 1989, Patek Philippe put an end to the quarrel with its Calibre 89. With 33 complications and 1,728 components, this timepiece currently stands as the most complicated portable watch in the world, though describing it as "portable" might be a stretch—at 88.2mm across and 41.07mm thick, weighing over 2 pounds, the Calibre 89 is not exactly a pocket or fob watch.

The advent of the wristwatch in the early 20th century obliged watchmakers to slightly restrict their ambitions for complicated pieces, given the smaller volume of the wristwatch's case. However, it also inspired them to come up with treasures of ingenuity to miniaturize existing complications and combine them in exceptional pieces. Today, the biggest names in the industry compete to create "the most complicated watch in the world," aided in their quest by computer software that allows for possibilities that were unimaginable just a few decades ago. In March of 2013, Marc Alexandre Hayek unveiled the prototype of a Breguet wristwatch offering 35 different functions on the piece's two faces. Named Hommage à Nicolas G. Hayek Ref. 7887, this piece has what it takes to become "the most complicated wristwatch in the world."

Pocket Records

Though pocket-watches have been largely replaced by wrist-watches, their larger cases provide horologers with a versatile field of experimentation when it comes to gathering as many complications as possible within the same timepiece. However, without shooting for the Guinness Book of World Records, larger horologers as well as artisans have recently offered extremely sophisticated pocket watches.

> **Vincent Bérard**
> *Quatre Saisons Carrosse*
> Artisan watchmaker Vincent Bérard has chosen to reinvent the carriage watch with the Quatre Saisons Carrosse. This handmade piece combines a perpetual calendar, quarter repeater, a system of auto-matons on the dial, a power reserve indicator and a thermometer.

> **Breguet**, *Modern Reproduction of the Marie-Antoinette*
> In 2008, Breguet presented an identical reproduction of the famous ultra-complicated Marie-Antoinette watch (1827), one of the most impressive achievements of the house's history, housed in a box carved from the queen's favorite oak at Versailles.

> **Patek Philippe**, *Calibre 89*
> With 33 complications and a 1,728-piece movement, the Calibre 89 from Patek Philippe is the most complicated portable watch in the history of horology. Besides the complete roster of "classic" complications (tourbillon, calendar, chronograph, strike, astronomical complications, etc.), it presents a few extremely rare functions such as the "secular" perpetual calendar (see chapter on Calendar watches), as well as a display of the date upon which Easter will next fall. It is also equipped with a celestial map that depicts the Milky Way and 2,800 stars of the Northern Hemisphere, as well as their relative size. Its development and production required nine years.

> **Patek Philippe**, *Star Caliber 2000*
> The double-sided pocket watch Star Caliber 2000, from Patek Philippe, associates 21 complications. Dedicated to our sun, the obverse dial displays functions including a running equation of time, times of sunrise and sunset, and a 24-hour day/night indication. The second dial, on the reverse side, pays tribute to the night sky with a map featuring the moon and stars, a moonphase display and a golden ellipse outlining the part of the sky visible from the selected observation site.

> **Cartier**
Grande Complication Squelette Pocket Watch

An ode to transparency, the Grand Complication Squelette Pocket Watch from Cartier associates a tourbillon, a monopushbutton chronograph, a perpetual calendar and an eight-day power reserve within its finely openworked 457-component movement.

< **Breguet**, *N° 5*

The N° 5 from Breguet, a replica of an older watch, associates automatic winding, 60-hour power reserve, moonphase display and a quarter repeater whose hammer strikes the inside of the case instead of a gong.

> **Audemars Piguet**, *Grande Complication Savonnette*

Within its yellow-gold case cover, the Grande Complication Savonnette (hunter-type watch) from Audemars Piguet unites a perpetual calendar, minute repeater and split-seconds chronograph.

Exploits on the Wrist

The race to produce "the most complicated wristwatch in the world" is in full swing. To stand out from the crowd, the most high-end watchmakers come up with feats of ingenuity, reinventing the base mechanisms that support the complications, most of which were invented and perfected before the 19th century. The time has come for new constructions, in new materials, with new interpretations that playfully overturn long-established traditions.

> **Patek Philippe**, *Sky Moon Tourbillon Ref. 6002*

The most complicated Patek Philippe wristwatch, the double-sided Sky Moon Tourbillon (2001) combines 12 complications, including a minute repeater with cathedral gongs, a perpetual calendar with retrograde date hand and a tourbillon. The dial on the reverse side presents a mobile map of the sky, reproducing—with extreme precision—the visible movement of the stars and the moon, as well as the moonphases. In 2013, this timepiece was presented in a new version paying tribute to fine handcraft professions and featuring an entirely hand-sculpted white-gold case framing a cloisonné and champlevé enameled dial.

∧ Dominique Loiseau, *1F4*

Produced in a limited edition of just two models per year, the 1F4 from Dominique Loiseau associates numerous functions, including a perpetual calendar, split-seconds chronograph, second time zone, equation of time, flying tourbillon, minute repeater and grand and small strike.

∧ Vacheron Constantin, *Tour de l'Ile*

In 2005, Vacheron Constantin celebrated its 250th anniversary by creating the ultra-complicated model called Tour de l'Ile. This double-sided timepiece contains in its 834-piece movement no fewer than 16 complications, including a minute repeater, tourbillon, indications of the moonphase and age of the moon, perpetual calendar, equation of time, times of sunrise and sunset, and a map of the night sky.

∧ Franck Muller, *Aeternitas Mega 4*

The Aeternitas Mega 4 model from Franck Muller houses, within its elegant tonneau-shaped case, 15 complications, including grand and small strike, minute repeater with Westminster chime, secular calendar with retrograde date, equation of time, large tourbillon, split-seconds chronograph and two additional time zones.

Hybris Mechanica à Grande Sonnerie

Hybris Mechanica à Gyrotourbillon

Reverso à Tryptique Hybris Mechanica

⌃ Jaeger-LeCoultre, *Hybris Mechanica*

The 2009 Hybris Mechanica trilogy from Jaeger-LeCoultre displays a total of 55 complications, some of which are found on more than one model. The Hybris Mechanica à Gyrotourbillon is an homage to the quintessence of the mechanical measure of time, thanks to its "spherical tourbillon" (see chapter on Tourbillons), completed by a perpetual calendar with four retrograde hands and an equation of time function that can be adjusted according to one's location. The Hybris Mechanica à Tryptique comes up with an unusual take on the famous Reverso reversible case, originally created in 1931. The watch boasts three faces, for three dimensions of time. On the front of the case is civil time, with a new detent escapement system and a titanium tourbillon; the back shows sidereal time, with a sky map, Zodiac calendar, equation of time and times of sunrise and sunset; while the watch cradle displays perpetual time, with a complete calendar—making a grand total of 18 complications on this exceptional timepiece. As for the Hybris Mechanica à Grande Sonnerie, it contains no fewer than 26 complications, with a 1,300-piece movement that powers a flying tourbillon, perpetual calendar with retrograde hands and a strike/minute repeater mechanism with numerous patented innovations.

Varied Triumphs

The horological superstars whose complications can be counted on the fingers of two, three, four, even five or six hands should not make us forget all the creations that—though they have achieved more modest recognition—are exceptional triumphs in their own right. A "grand complication," in the watch world, is created when a watch contains three or four supplementary functions—especially if each of those functions belongs to one of the traditional domains of haute horology (audible indications, calendars, chronographs, tourbillons, etc).

Cartier
Rotonde Grand Complication Skeleton

Issued in a limited edition of 30 pieces, Cartier's Rotonde Grand Complication Skeleton boasts a tourbillon, perpetual calendar, single-pusher chronograph and eight-day power reserve display—all within an entirely openworked, decorated 457-piece movement.

Blancpain
Le Brassus Carrousel Répétition Minutes Chronographe Flyback

On the Le Brassus Carrousel Répétition Minutes Chronographe Flyback model from Blancpain, the openworked dial and sapphire exhibition caseback provide a fine opportunity to admire the triple complication in action—and notably the famous carrousel (or karussel), an alternative to the tourbillon that was revived by the brand in 2008.

IWC
Portuguese Grand Complication

The Portuguese Grand Complication from IWC offers a minute repeater and chronograph, as well as a perpetual calendar that indicates not only the date, day and month, but the year as well, via a four-digit display.

A. Lange & Söhne
Grande Complication

The Grand Complication model is the most sophisticated watch from A. Lange & Söhne and combines a large and small strike (with minute repeater), a perpetual calendar and a split-second chronograph with jumping seconds. It is issued in a six-piece limited edition.

Jean Dunand
Grande Complication

In 2005, the brand Jean Dunand—co-founded by Christophe Claret, the renowned constructor of complicated movements—presented a wristwatch called Grande Complication. This 827-piece timepiece includes, among other functions, a single-pusher split-seconds chronograph, tourbillon, minute repeater and perpetual calendar with retrograde date and day indications.

Audemars Piguet
Royal Oak Offshore
Grande Complication

Equipped with highly visible architecture that reveals its openworked movement, the Royal Oak Offshore Grande Complication brings together a perpetual calendar, minute repeater and split-seconds chronograph within its titanium and ceramic case.

Vacheron Constantin
Patrimony Traditionnelle Calibre 2755

A worthy descendant of the prestigious Tour de l'Ile, released for the 250th anniversary of the brand, the Patrimony Traditionnelle Calibre 2755 from Vacheron Constantin brings together three major complications of the horological universe—minute repeater, tourbillon and perpetual calendar—in an elegant round case.

Maîtres du Temps, *Chapter One*

Presented in 2008 in a tonneau-shaped case and rereleased in 2010 in a round version, Chapter One from Maîtres du Temps unites a single-pusher column-wheel chronograph, a retrograde second time zone display and retrograde date, as well as day and moonphase displays on two rollers at 6:00 and 12:00.

Jean Dunand, *Shabaka*

Boasting a highly original, Art Deco-inspired dial, Shabaka from Jean Dunand associates a minute repeater with cathedral chime with a complete perpetual calendar with instantaneous day/date/month displays on rollers.

Christophe Claret, *DualTow NightEagle*

Christophe Claret scored a huge coup with its DualTow NightEagle, which features hour/minute displays on two belt drives inspired by treaded vehicles, a single-pusher planetary chronograph with strike and tourbillon, not to mention many other refinements and technical innovations.

HD3, *Slyde*

An ultra-modern, ultra-geeky version of grand complications, the new electronic model Slyde from HD3 is distinguished by its curved tactile screen that displays the various indications (time, calendar, chronograph, countdown clock, moonphase, second time zone) in much the same way as a smartphone.

Patek Philippe
Triple Complication Ref. 5208

The automatic Triple Complication Ref. 5208 wristwatch from Patek Philippe associates within its platinum case a minute repeater, single-pusher chronograph and instantaneous perpetual calendar with apertures. It also stands out for its escapement and balance spring in Silinvar®, a silicon derivative.

Patek Philippe, *Ref. 5216*

Reinterpreting one of the manufacture's classics, Patek Philippe's Ref. 5216 conceals beneath its sober dial a minute repeater, tourbillon, perpetual calendar with retrograde date hand, and a moonphase display.

Audemars Piguet
Tradition Minute Repeater Tourbillon Chronograph

The Tradition Minute Repeater Tourbillon Chronograph by Audemars Piguet accommodates its three complications within a cushion-shaped pink-gold or titanium case topped by a white-gold bezel. A transparent exhibition caseback provides a chance to admire the movement decorations.

A. Lange & Söhne, *Lange 1 Tourbillon Quantième Perpétuel*

The Lange 1 Tourbillon Quantième Perpétuel from A. Lange & Söhne associates its two classic complications in a design that is unmistakably recognizable as the Lange 1, with its off-centered hours/minutes display and a large date in a double aperture.

Astronomical Watches

Astronomical complications allude to the very source of the art of horology, and they are part of the classic repertory of all ultra-complicated pieces. At the start of the third millennium, they are making a forceful comeback on highly sophisticated pieces that meld technical prowess and poetry. The major rising star is the map of the night sky.

IWC, *Portuguese Sidérale Scafusia*

The most complex watch ever created by IWC, the Portuguese Sidérale Scafusia displays on its double-sided case the solar time, sidereal time (that of the stars) and various other astronomical indications including a map of the night sky and times of sunrise and sunset—not to mention a constant-force tourbillon and perpetual calendar.

Patek Philippe, *Celestial Ref. 5102*

The Celestial Ref. 5102 from Patek Philippe continuously displays the exact configuration of the night sky in the Northern Hemisphere, with the apparent movement of the stars, the position of the moon and its phases over the course of the lunar cycle.

Audemars Piguet, *Royal Oak Equation of Time*

Equipped with an equation of time system that takes longitude into account (see chapter on Equations of Time) the Royal Oak Equation of Time from Audemars Piguet is also one of the rare wristwatches to display times of sunrise and sunset utilizing three parameters: date, longitude and latitude.

Louis Moinet, *Astralis*

Besides its astral tourbillon and split-seconds column-wheel chronograph, the Astralis from Louis Moinet—a limited edition of just 12 pieces—stands out for its 24-hour planetary function, which successively displays four planets, each composed of an extremely rare meteorite.

Van Cleef & Arpels, *Midnight in Paris*

The Midnight in Paris model by Van Cleef & Arpels reproduces on its aventurine dial the movement of the stars in the sky above Paris, the City of Light. The disc around the calendar on the caseback is set with meteorite.

Jaeger-LeCoultre, *Master Grande Tradition Grande Complication*

On the Master Grande Tradition Grande Complication from Jaeger-LeCoultre, the "orbital" flying tourbillon that serves as an hour hand indicates sidereal time, or time calculated by the stars. Civil time is indicated by a small sun turning around the periphery of the dial, and a Zodiac calendar indicates the position of the constellations as they appear throughout the year.

Chopard, *L.U.C All-in-One*

To celebrate its 150th anniversary, Chopard presented the L.U.C All-in-One model. Possessing 516 components, the L.U.C 4TQE movement—certified as a chronometer and marked with the Geneva Seal—displays 14 complications (tourbillon, perpetual calendar, etc.) on the front and back of the watch, including, on the caseback, equation of time, power reserve, 24-hour day/night indication, times of sunrise and sunset and astronomical moonphase.

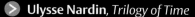

Ulysse Nardin, *Trilogy of Time*

Ulysse Nardin has demonstrated its commitment to the astronomical and astrological science of the Renaissance by creating an exceptional trio of timepieces called Trilogy of Time. The first piece, Astrolabium Galileo Galilei, depicts on its dial the functions of the astrolabium, a navigational tool that facilitated great discoveries. It indicates the respective positions of the sun, moon and stars as observed from the Earth. The Planetarium Copernicus continuously displays the month, Zodiac sign and astronomical positions of the planets in relation to the sun and the Earth using a dial composed of seven concentric rings. As for the Tellurium Johannes Kepler, with its cloisonné enamel dial representing the Earth as seen from above the North Pole, it displays times of sunrise and sunset (with areas of shadow and light), as well as solar and lunar eclipses.

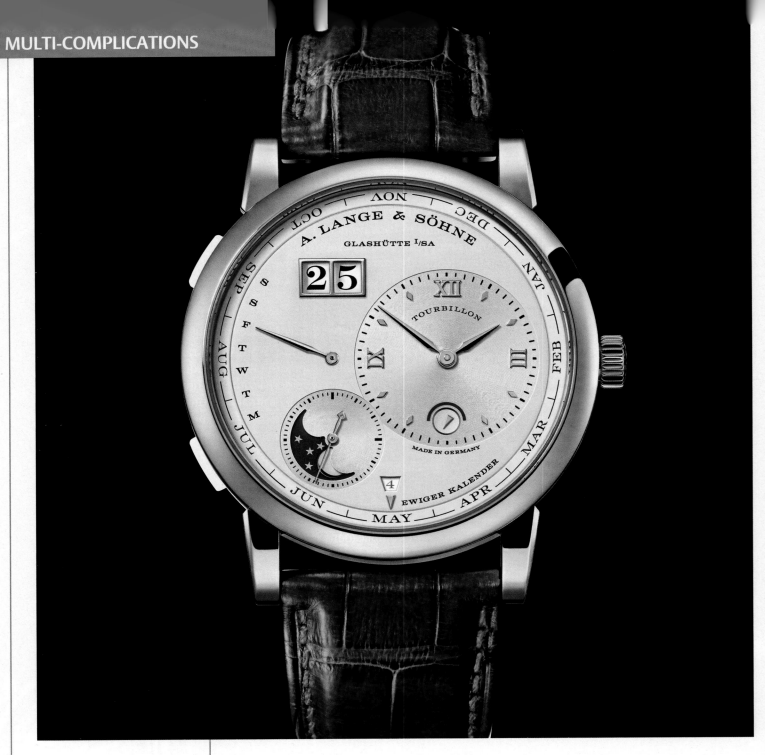

A. LANGE & SÖHNE

LANGE 1 TOURBILLON PERPETUAL CALENDAR – REF. 720.032

The perpetual calendar and moonphase displays are enhanced by the manufacture's characteristic silver-toned guilloché pattern on this classically understated dial. The contemporary architecture of the DB2509 caliber beats at a rate of 36,000 vph, driving one of the fastest and lightest tourbillons (comprising 63 parts weighing a total 0.18g). This movement is also equipped with a power reserve indication, a seconds indication, and a retrograde moonphase indicator that facilitates setting the moon to the exact date.

A. LANGE & SÖHNE

RICHARD LANGE PERPETUAL CALENDAR "TERRALUNA" – REF. 180.026

This hand-wound wristwatch houses its 787-component L096.1 caliber in a 45.5mm white-gold case. The movement, endowed with a sophisticated constant-force escapement that ensures unvarying amplitude throughout the 14-day power reserve, animates a perpetual calendar of ingenious architecture. Told on three separate axes, the hours, minutes and seconds are met in their respective subdials by apertures displaying the day, month and large date. A discreet leap-year indicator at 2:30 completes the dial-side exhibition. On the caseback, the Terraluna's orbital moonphase display depicts the moon orbiting the earth once per precise synodic month while the moon's phases and relative position are illustrated via a constellation view that incorporates the balance as a representation of the sun.

A. LANGE & SÖHNE

SAXONIA ANNUAL CALENDAR – REF. 330.025

The platinum version of the Saxonia Annual Calendar lends a sophisticated shine to a model that became an instant classic upon its 2010 release. The piece's calendar indication includes the day, date, month and moonphase. Taking into account the variance between months of 30 and 31 days, the calendar mechanism need only be corrected at the end of February. Its prominent moonphase indication loses only one day every 122 years. A peek through the sapphire crystal caseback reveals the automatic-winding manufacture caliber L085.1 SAX-O-MAT, constructed to show off the classic screw balance—the watch's beating heart.

AUDEMARS PIGUET

JULES AUDEMARS OPEN-WORKED TOURBILLON CHRONOGRAPH
REF. 26353PT.OO.D028CR.01

The Jules Audemars Open-Worked Tourbillon Chronograph is powered by Audemars Piguet's in-house Manufacture caliber 2889, which is comprised of 290 parts and beats at 21,600 vph. This timepiece has a platinum case featuring a fixed bezel, lugs, a faceted crown and two pushbuttons at 2:00 and 4:00. It fascinates by the sheer complexity of its open-worked dial, which bears a tourbillon carriage at 6:00 as well as subdials. Its two blued steel hands and a slender sweep seconds hand glide around the blue Arabic numerals appearing on the inner bezel ring. The strap is large-scale alligator leather.

AUDEMARS PIGUET

ROYAL OAK OFFSHORE GRANDE COMPLICATION -- REF. 26065IS.OO.11051S.01

The Royal Oak Grande Complication is powered by the 2885 caliber, a self-winding movement of 13.5''', which features a split-seconds chronograph, a minute repeater and a perpetual calendar. Seconds are displayed in a subdial at 9:00. Housed in an 18K white-gold case, the movement beats at 19,800 vph. The dial displays 12 indications, including day, date, month, leap years and moonphase. The movement is comprised of 654 pieces.

AUDEMARS PIGUET

TRADITION MINUTE REPEATER TOURBILLON CHRONOGRAPH
REF. 26564IC.OO.D002CR.01

The Tradition Minute Repeater Tourbillon Chronograph houses the hand-wound, in-house Manufacture caliber 2874, which is enclosed in a titanium case with a white-gold bezel, crown, pushbuttons and caseback. The dial is silvered opaline and has pink-gold applied Arabic numerals and hands. The movement is comprised of 504 parts and beats at 21,600 vph. Its functions include a minute repeater, tourbillon, chronograph, hours, minutes and small seconds. The minimum power reserve is 48 hours. This model is available in a limited edition of 10 pieces.

CHRISTOPHE CLARET

21 BLACKJACK – REF. MTR.BLJ08.340-361

A veritable miniature casino, the 21 Blackjack matches grand complications with the world of gaming, in the process creating a new watchmaking paradigm: the interactive watch. Rather than just tell the time, the wearer can enjoy the sensory effects of blackjack, roulette and dice. On the lower part of the dial, the player's four cards appear in windows. Two are visible, the other two hidden by shutters. On the upper part of the dial are three additional windows for the dealer's cards, one of which is visible, the other two also hidden by shutters. The 21 Blackjack is an unprecedented upmarket timepiece for aficionados, initiating a kind of watchmaking that has cast off its inhibitions.

CHRISTOPHE CLARET

BACCARA – REF. MTR.BCR09.020-029

Baccara is a timepiece doubling as a miniature casino with three games: baccara, roulette and dice, each appealing to the audio, visual and tactile senses. The Baccara cards appear on the dial as if by magic. At 6:00, the player's cards are distributed in three small windows. The banker's cards are at 12:00. To shuffle the cards, the player presses a pusher located at 9:00 while a pusher at 8:00 distributes the cards to the players. At 10:00, a third pusher organizes distribution to the bank. Every time the shutter opens, for either the player or the bank, a cathedral gong can be seen through a caseband window at 2:00.

CHRISTOPHE CLARET

POKER – REF. MTR.PCK05.001-020

Poker is the very first timepiece that manages to replicate the card game in an automaton watch. While the game seems simple to organize on a table, its watchmaking solution is an extreme test and no technical solution had been found until Mr. Christophe Claret himself came up with the answer. While the first prototype was working in 2011, it took two years to perfect the intricate complication. In total, Poker packs 32,768 different combinations, i.e. 98,304 combinations for three players. The probabilities were calculated so that each player has approximately the same chances of winning. The original automatic-winding PCK05 caliber comprises 655 components and has a power reserve of 72-hours. There are bound to be some late nights! In addition to its gaming functions, Poker has an extremely legible display of hours and minutes.

CHRISTOPHE CLARET

MAESTOSO – REF. MTR.DTC07.020-028

With Maestoso, Christophe Claret took up the challenge of incorporating a traditional pivoted detent escapement, usually designed to run in a perfectly stabilized position, in a wristwatch. Enabling the detent escapement to operate in all the positions that wearing a wristwatch entails is a genuine gamble, pulled off via three patented mechanisms and a series of additional, innovative systems. To prevent the detent from turning over, an anti-pivot cam, integral to the spring balance, works in conjunction with the safety finger. The ensemble is fitted between a mainplate and two sapphire bridges and pivots on a ball bearing, a patented mechanism, that distributes the load on the escapement so it de facto absorbs impacts by means of a spring, which gives it the requisite flexibility.

DE BETHUNE

DB16 PERPETUAL CALENDAR DEADBEAT SECONDS TOURBILLON

The perpetual calendar and moonphase displays are enhanced by the manufacture's characteristic silver-toned guilloché pattern on this classically understated dial. Enthralled by the soft, steady beat of the jumping seconds, the wearer can sense the very core of horological wonderment as they experience a unique form of ultimate and uncopromising beauty. The contemporary architecture of the DB2509 caliber beats at a rate of 36,000 vph, one of the fastest and lightest tourbillons, comprising 63 parts weighing a total 0.18g. This movement is also equipped with a power indication, a seconds indication, and a retrograde moonphase indicator that facilitates setting the moon to the exact date.

DE BETHUNE

DB25 LT TOURBILLON

As if literally floating in space, the spherical De Bethune moon sits delicately at 12:00 and opens a symbolic window onto the cosmos. Entirely mirror-polished and composed of half-spheres meticulously set into blued steel and platinum, this three-dimensional indicator is driven by a high-precision mechanism requiring only one adjustment every 1,112 years. Alongside classic features such as a triangular bridge adorned with Côtes de Bethune, the back of the DB25 LT displays a strong contrast with its avant-garde mechanism powered by a 30-second silicon and titanium tourbillon with its balance-wheel oscillating at 36,000 vph.

DE BETHUNE

DB25 T DEADBEAT SECONDS TOURBILLON

Thanks to new technologies, De Bethune has created a 0.18 g silicon-titanium tourbillon in a carriage spinning once every 30 seconds on its axis, and a balance oscillating at a frequency of 36,000 vph. This tourbillon is four times lighter than its classic counterparts. In the style of regulator timepieces, this model has a jumping seconds display at the center of the watch, along with a double lever with four pallets to orchestrate the gold double wheel of this mechanism. The result is an impressive mechanical ballet.

DE BETHUNE

DB28 ST DEADBEAT SECONDS TOURBILLON

The DB28 ST is a contemporary expression of traditional mechanical horology. Each aesthetic detail has its own specific reason for being. The soft gleam of the platinum bezel effects the transition from the cool shade of the titanium case to the warmth of the sterling silver chapter ring. The layered dial structure is built around suspension bridges featuring fine lines and multiple decorations serving to elevate the central mechanism as well as enhancing its functionality. The silicon/titanium tourbillon and the jumping seconds mechanism are proudly displayed at the heart of this model, endowing it with peerless technical and aesthetic splendor.

FRANCK MULLER

AETERNITAS MEGA

Simply put: the most complicated wristwatch ever made! With 25 of this masterpiece's 36 complications visible to the wearer's amazement, the Aeternitas Mega may make its most impressive claim in the clarity with which its multitude of complexities are displayed. With an array of retrograde counters, a highly intricate single-pusher split-seconds chronograph, an extraordinary eternal perpetual calendar, the choice of grand or small strike of the Westminster Cathedral carillon, and two additional 24-hour time zones, to name a few of the 36 modules, this 1483-component magnum opus demonstrates the infinity of Franck Muller's daring vision of haute horology.

GIRARD-PERREGAUX

GIRARD-PERREGAUX CONSTANT ESCAPEMENT LM – REF. 93500-53-131-BA6C

A true technical revolution that stunned connoisseurs when prototypes of the innovative mechanism were first presented, the Constant Escapement is now integrated into the movements driving new models of Girard-Perregaux's Haute Horlogerie collection. Based on groundbreaking prototypes presented in 2008, it has taken five years of research and development to bring this signature movement to fruition. Unlike other systems that propose a constant force averaged over time, this is an authentic constant-force escapement, as the force is both instantaneous and continuous. This is the first model to house this innovative and original escapement and it is garbed in resolutely technical and contemporary design.

GLASHÜTTE ORIGINAL

PANOLUNAR TOURBILLON – REF. 93-02-05-05-05

Housed in a 40mm red-gold case with a warm silver dial, the characteristic asymmetric design of the PanoLunar Tourbillon finds the large hour and minute counter positioned to the left of the center, in alignment with the flying tourbillon set within it and centered at 7:00. The Panorama date window to the lower right presents black numerals on an ivory ground, perfectly matched with the warm silver dial. To the upper right, the moonphase shows a golden moon and shimmering stars against a dark blue night sky.

HUBLOT

MP-02 KEY OF TIME – REF. 902.NX.1179.RX

Hublot builds a time machine... and relinquishes control to its wearer. Armed with a tri-positional crown, this micro-blasted titanium wristwatch permits its owner to modulate the speed of passing time. While position 1 slows the movement of the hands by a factor of four, representing one hour as a quarter thereof on the dial, position 3 does the opposite, accelerating the choreography to display 15 minutes as a full rotation. Made possible by the 512-component movement's genuine "mechanical memory," a return to position 2 sets the hands back to a conventional time display, concealing the time travel from unwanted eyes. Seconds are indicated on the top edge of the watch's vertical flying tourbillon at 6:00.

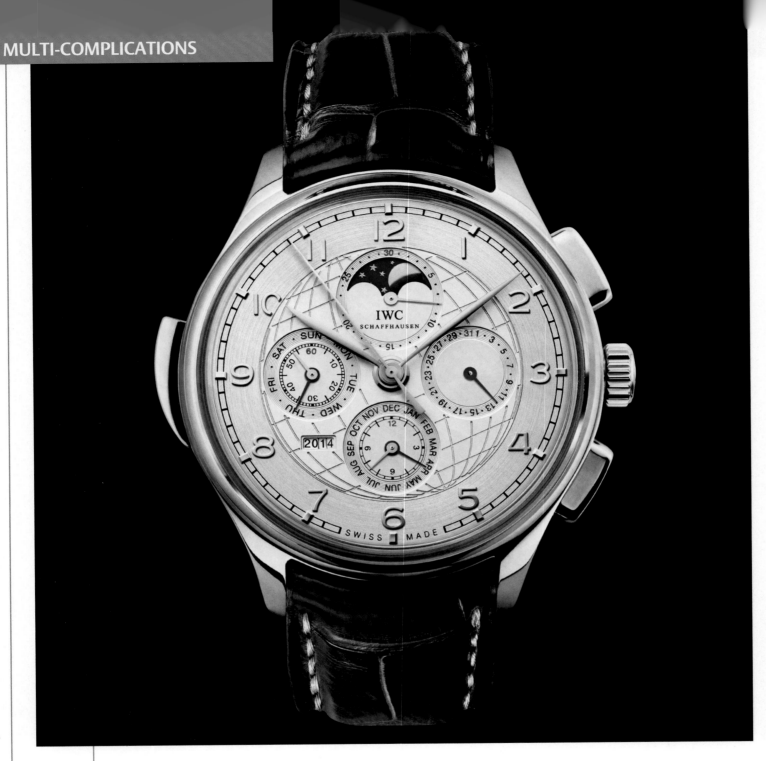

IWC

PORTUGUESE GRAND COMPLICATION – REF. IW377401

The Portuguese Grand Complication combines a multiplicity of extraordinary horological feats within a platinum case. The cool color of the case and black alligator leather strap, stitched with platinum threads, draws the eye to a cartographic white dial design that pays tribute to Vasco da Gama's navigational exploits. Along with a full perpetual calendar, programmed until the year 2499, and a small seconds subdial, this flagship masterpiece features a chronograph with central seconds hand, as well as 12-hour and 30-minute counters, a perpetual moonphase display and a minute repeater. On the outside of the case, the crown, chronograph pushers and repeater's sliding activator echo the brilliance of the dial's numerous appliqués. The theme of navigational exploration extends to the caseback, upon which a sextant is engraved.

JACOB & CO.

PALATIAL FLYING TOURBILLON MINUTE REPEATER – REF. 150.500.24.NS.OB.1NS

Beneath a semi-transparent dial in blue mineral crystal, this limited edition wristwatch's hand-wound JCBM03 caliber drives a sophisticated timekeeping choreography. Boasting a titanium balance, one-minute flying tourbillon visible through the face at 6:00, and power reserve of 90 hours, the 308-component movement powers the sonorous delight of a minute repeater, activated via an 18K rose-gold sliding lever at 9:00. The titanium case, measuring 43mm, provides a luxurious contrast to the dial's blued tones with its adornment of a crown in 18-karat rose gold.

JAEGER-LECOULTRE

DUOMETRE SPHEROTOURBILLON – REF. 6052520

An authentic revolution in the field of Grand Complication models, the Duomètre Sphérotourbillon is equipped with an unprecedented multi-axis tourbillon. In addition to the axis of its carriage, the Sphérotourbillon has a second axis inclined at 20° and thereby freeing it from the effects of gravity in all positions. The revolutionary Dual-Wing concept featured in Calibre 382 once again paves the way for an original function and makes this exceptional model the first tourbillon watch that is precisely adjustable thanks to the small seconds flyback system.

JAEGER-LECOULTRE

MASTER GRANDE TRADITION GRANDE COMPLICATION – REF. 5023580

With the creation of this multi-complication watch in the Master Grande Tradition line, Jaeger-LeCoultre has once again demonstrated its profound attachment to its historical roots, while expressing its ambition to adapt watchmaking expertise to the latest research developments. To convey the perfect blend of haute horlogerie and advanced technology, this piece houses a flying tourbillon that is cantilevered and thus held only at the underside in an ultra-light titanium cage. The cage rests at 6:00 on a beautiful dial, which represents a canopy of heavens. This model is not only a treat to the eyes, but to the ears as well, as it houses a musically innovative minute repeater.

JAEGER-LECOULTRE

MASTER GRANDE TRADITION GYROTOURBILLON 3 JUBILEE – REF. 5036420

With an aesthetic inspired by the pocket-watches of the 19th century, this is the third Jaeger-LeCoultre complication to be equipped with a Gyrotourbillon, but the first with a flying Gyrotourbillon. The manufacture introduces a flying tourbillon watch whose absence of upper bridge allows it to overlook the movement exercised by this miniature universe. At the heart of two extra-bright aluminium cages, a 14K gold balance wheel is visible as well as a new blued spherical balance spring, as if beating like a heart at the center of the Gyrotourbillon. Fitted with a new spherical balance-spring, it is also equipped with a chronograph with an instantaneous digital display activated by a monopusher. This timepiece with exceptional finishes reflects the Manufacture's inventiveness.

JAEGER-LECOULTRE

MASTER GRANDE TRADITION TOURBILLON CYLINDRIQUE A QUANTIEME PERPETUEL JUBILEE – REF. 5046520

This timepiece is a perfect combination of the historic heritage of 19th-century watches and cutting-edge technology. Its perpetual calendar indications are displayed extremely clearly on its dial. The day and moonphase are at 3:00, the month and year are at 12:00 and the date is at 9:00. This model is the first to combine a flying tourbillon with a cylindrical balance-spring, thus guaranteeing an exceptional level of timekeeping performance. All in all, this is a timepiece of great precision, combining exceptional complications with an extremely expressive aesthetic appearance.

PANERAI

LO SCIENZIATO LUMINOR 1950 TOURBILLON GMT CERAMICA – 48MM – REF. PAM00528

Lo Scienzato is one of the many sobriquets earned by Galileo Galilei due to his devotion to research that extended the boundaries of understanding through scientific learning. His studies had a profound impact on mechanical timekeeping, which would not exist today without his formulation of the law of the pendulum. This model pays tribute to Galileo as it is a piece of remarkable technical mastery. The skeletonized piece offers its owner full view of every internal detail, including the tourbillon cage, which makes two rotations per minute on an axis at a right angle to that of the balance, providing the best possible compensation for errors of rate caused by gravity. Additionally the second time zone indicator sits at 3:00.

TECNICA PALME – REF. PFH435-1207000-HA1441

This unique hand-wound masterpiece in 18K white gold adorns its hand-engraved white-gold dial with the majestic visual beauty of a translucent enamel covering. Visible through the colorful stage, the 578-component PF 352 caliber's intricate tourbillon brings a depth of dimension to a comprehensive display equipped with an ingenious twist of horological engineering. Joining the enameled plique-à-jour hour and minute hands, the timepiece's perpetual calendar is efficiently presented via three subdials that perfectly extend the smooth curvature of the vibrant décor. The Tecnica Palme's bezel cleverly doubles as a winding mechanism for the movement's sonorous two-cathedral-gong minute repeater. A chronograph completes the ultra-sophisticated yet harmonious architecture.

PARMIGIANI

TONDAGRAPHE TOURBILLON REF. PFH236-1201400

Accentuated by a superb juxtaposition of layered components on an opaline dial with circular snailed pattern, this expressive timepiece extends its multi-dimensional architecture all the way to its dynamic heartbeat. Complementing the hand-wound PF 354 caliber's central-seconds chronograph with 30-minute counter at 3:00 and retrograde power reserve indicator at 12:00, the watch's stunning tourbillon at 6:00 affords the wearer a detailed view of its meticulous construction. Housed in a 43mm 18K white-gold case, this 10-piece limited edition is individually numbered and worn on an Hermès alligator strap with 18K gold ardillon buckle.

PATEK PHILIPPE

MEN COMPLICATIONS – REF. 5960P-016

Housed in a 40.5mm platinum case, this masculine timepiece combines a chronograph and annual calendar on a highly legible black opaline dial with white-gold hour markers. While the lower half of the face is devoted to the central chronograph hand's complementary functions, in the form of a dual-purpose 12-hour/60-minute counter with subtly integrated day/night indicator, the upper section displays the day, date and month through three windows between 10:00 and 2:00. The 456-component self-winding CH 28-520 IRM QA 24H caliber, with Breguet spiral and Gyromax balance, boasts a power reserve of 45-55 hours, visible below the date indication.

PATEK PHILIPPE

MEN GRAND COMPLICATIONS – REF. 5160R-001

When uncompromising finesse, meticulous attention to detail and unparalleled mechanical sophistication meet on a single wristwatch, a masterpiece is born. Within the exquisite intricacy of a 38mm rose-gold case engraved entirely by hand, this self-winding timepiece's white opaline dial, with hand-engraved central motif, unites the classical delicacy of 10 black Roman numerals and the modernity of its state-of-the-art 324 S QR movement. Complete with moonphase at 6:00, and day, month and leap-year indications via apertures at 9:00, 12:00 and 3:00, this timepiece's perpetual calendar surrounds the splendid interior décor with a hand-guided retrograde date display. Turning over the watch, the sapphire crystal caseback is protected by a hinged cover.

PIAGET

PIAGET GOUVERNEUR TOURBILLON – REF. G0A37115

A member of Piaget's Black Tie collection, the Gouverneur Tourbillon combines elegant aesthetics with technical expertise. Entirely manufactured in house, this model is equipped with the ultra-thin, hand-wound 642P caliber beating at 21,600 vph with a power reserve of 40 hours. The silvered guilloché dial showcases a tourbillon with small seconds on the cage at 12 as well as an astronomic moonphase at 6:00. The silvery, cool color of the dial is complemented by the 18K white-gold case, which is perfectly set with two rows of diamonds (1.4 carats). This model possesses two of the grand complications all while exhibiting elegance and beauty.

ROGER DUBUIS

EXCALIBUR QUATUOR IN PINK GOLD – REF. RDDBEX0367

The Excalibur Quatuor watch is produced in a limited series. It represents innovation while respecting the great traditions of watchmaking by incorporating four balance springs as well as a completely original power reserve indicator. These two technical feats have not been developed for their own sake but as part of watchmaking's permanent pursuit of precision. The precision of a movement is affected by gravity of the Earth as the watch constantly changes position with the motion of the wrist. While the tourbillon has provided a partial answer to this problem, the Excalibur Quatuor offers a new solution, with its four carefully positioned sprung balances working in pairs. This immediately compensates for the rate variations caused by changes in position.

TAG HEUER

CARRERA MIKROPENDULUMS – REF. CAR2B81.FC6323

An exclusive TAG Heuer patented technology: the Carrera MikropendulumS contains the first magnetic double tourbillon. Equipped with two magnetic pendulums that replace traditional hairsprings, this chronograph tourbillon rotates 12 times per minute. The two tourbillon pendulums and their solid 5N 18K rose-gold bridges are visible through the fine-brushed anthracite dial. The case is forged from a revolutionary material, a chrome and cobalt alloy used in aviation and surgery. It is fully biocompatible, harder than titanium, easier to shape and as luminous as white gold. The case design, with its stopwatch-like placement of the crown at 12:00, is based on the 2012 Aiguille d'Or winner, the TAG Heuer Carrera Mikrogirder, and the Carrera 50 Years Anniversary Jack Heuer edition.

ZENITH

ACADEMY CHRISTOPHE COLOMB HURRICANE – REF. 18.2212.8805/36.C713

A combination of enameled and guilloché dials within a multi-layered architecture gives this 45mm 18K rose-gold timepiece an air of fascinating depth and three-dimensional weightlessness. On the upper half of the face, the off-centered hours and minutes are displayed on an open subdial that reveals the El Primero 8805 caliber's exceptional constant-force fusée-and-chain transmission. On the lower half, under a domed sapphire crystal, the 939-component movement's self-regulating gravity-control module is showcased as it ensures the perfect horizontal position of the regulating organ via a unique gyroscopic system.

ZENITH

ACADEMY CHRISTOPHE COLOMB HURRICANE GRAND VOYAGE – REF. 18.2211.8805/36.C713

A revolutionary display of pioneering innovation, this 18K rose-gold masterpiece showcases a patented module that ensures the perfect horizontal positioning of the regulating organ, regardless of the movements of the wearer's wrist. The riveting frontal architecture, highlighted by the aforementioned gyroscopic mechanism at 6:00 and constant-force fusée-and-chain transmission beneath the hours and minutes at 12:00, is met on the caseback by stunning display of miniaturized artistry. Depicting the Santa Maria, Christopher Columbus's 1492 vessel, as well as a sextant and an intricately engraved portrait of the explorer, the highly detailed, hand-crafted décor was achieved through an wealth of exceptional techniques, including enameling, champlevé and micro-engraving.

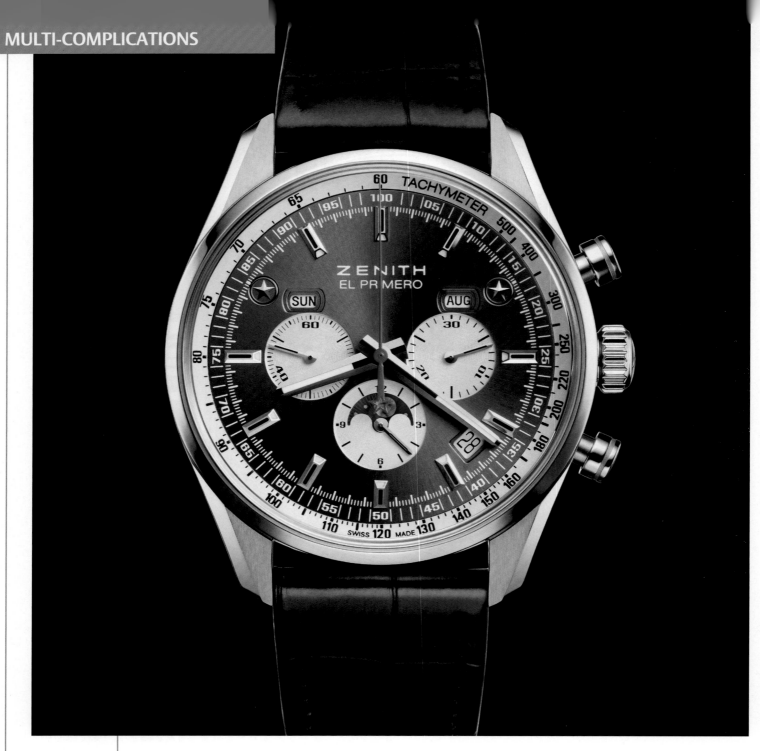

ZENITH

EL PRIMERO 410 – REF. 27.00.2118.496

The stainless steel self-winding wristwatch combines the historic excellence of Zenith's El Primero movements with a full calendar display and moonphase indication. Powered by the 36,000 vph El Primero 410 caliber with 50-hour power reserve, this 42mm timepiece invigorates its gray-toned silver sunray dial with the combination of a red central chronograph hand, apertures for day, date and month, small seconds at 9:00, 30-minute counter at 3:00 and a dual-purpose 12-hour totalizer/ moonphase subdial at 6:00. A tachometric scale, permitting the measure of speed or distance when used against the high-precision chronograph, surrounds the comprehensive architecture.

EL PRIMERO CHRONOMASTER GRANDE DATE – REF. 27.00.22.2218.713

Boasting a colorful display of the moon's and sun's phases on a subdial at 6:00, this self-winding 18K rose-gold and stainless steel timepiece uses the full breadth of its 45mm diameter to invigorate the measure of time. Revealed through an opening in the upper left quadrant of the silver-toned sunray dial, the legendary heartbeat of Zenith's El Primero movement, with silicon escape wheel and lever, adds a dynamic depth of dimension to this sophisticated creation. The large date is presented at 2:00, above the column-wheel chronograph's 30-minute counter.

Equations of Time

Blancpain
Villeret Equation de Temps Marchante

The 2004 launch

of the first "running" equation of time wristwatch

briefly made people believe in a revival of this very ancient complication. Ten years later, one is forced to admit that apart from a few models, generally emanating from the most prestigious manufacturers, this function remains the exclusive preserve of certain ultra-complicated watches. It is certainly true that displaying the difference between mean solar time (that shown on our watches) and solar time (that of nature and sundials) is a somewhat abstract notion for most of us. Yet it adds a little touch of poetry to dials, reminding us that our first clock is the sun and thereby renewing immemorial ties with Mother Nature.

THERE ARE ACTUALLY TWO TIMES:

true solar time and mean solar time. True solar time, which you might see on a sundial, corresponds to natural rhythms. Earth describes an ellipse in turning around the sun, and its axis is inclined in relation to its orbit. This means that the lapse of time between two passages of the sun through its highest point in the sky (noon) is not the same length year-round. It actually lasts 24 hours only four times a year: on April 15th, June 14th, September 1st and December 24th. Otherwise, it is sometimes longer and sometimes shorter, following an immutable curve. This difference ranges from 16 minutes and 23 seconds, on November 4th, to 12 minutes and 22 seconds, on February 11th and is called the "equation of time." As for mean solar time (also called civil time), it is a convention based on the average of all days in a year.

Since the famous Astrarium clock by Giovanni de' Dondi (1364), the greatest watchmakers have rivaled each other in ingenuity to develop systems that reproduce the variations in the equation of time, first in clocks, then, since the turn of the 18th century, in wristwatches. This function has become an indispensable feature of all ultra-complicated watches—particularly the astronomical ones—but it is much more rare to find it flying solo. In 2004, the "running" equation of time (see "How does it work?"), long reserved for pocket-watches, made its wristwatch debut.

HOW DOES IT WORK?

Since the variable lengths of true solar days are identically repeated on the same days, they can be "programmed" by means of a so-called "equation" cam that performs one full rotation per year. Its profile resembles that of an "analemma," representing the figure eight traced in the sky by the sun's various positions as recorded each day at the same time and from the same place throughout the calendar year. The extreme precision of this shape determines the accuracy of the equation of time.

DISPLAY METHODS

The equation of time display may be presented in various ways. Most watches are equipped with a subdial or auxiliary segment swept over by a hand running from -16 to +14 minutes (sometimes from -15 to +15); to know true time, the user must then do a mental addition or subtraction of this difference in relation to mean time.

RUNNING EQUATIONS

Distinctly more user-friendly yet far more complex in terms of their construction, "running" equation of time watches have two hands fitted coaxially with the minute hand, respectively indicating mean time and true time. They are equipped with a sensor system that searches out the information from the cam, as well as a differential that corrects the running of the supplementary minute hand. They require no mental arithmetic, since true time is continuously displayed directly on the dial.

A Very Rare Complication

Extremely rare as a solo complication, the equation of time is most often joined by a perpetual calendar, even an annual calendar, or with astronomical indications. It is usually displayed on a small fan-shaped segment covered by a hand, but certain watchmakers prefer to innovate by spotlighting the complication with a central hand or opting for a linear indication.

▶ Zenith
Academy Christophe Colomb
Equation of Time

Equipped with an escapement that maintains a horizontal position, whatever the position of the watch—as with the gimbals used in maritime navigation—the Academy Christophe Colomb Equation of Time from Zenith displays the equation of time across a fan-shaped segment at 9 o'clock.

Girard-Perregaux
1966 Annual Calendar and Equation of Time

The 1966 Annual Calendar and Equation of Time by Girard-Perregaux is distinguished by its dial, on which an extremely original configuration features the date in a small subdial at 1:30, the small seconds at 9 o'clock, the month in a panoramic window at 7 o'clock and the equation of time on the arc of a circle at 4:30.

Breguet
Classique Complications 3477BA/1E/986

On its finely guilloché dial, this grand complication watch from Breguet's Classique collection combines a perpetual calendar, power reserve display and an equation of time indication in a fan shape at 1:30.

Chopard, *L.U.C 150 All-in-One*

Chopard uses the back of its anniversary model L.U.C 150 All-in-One as a second dial: it features a power reserve display, 24-hour night and day indication, sunrise and sunset times, an orbital moonphase and an equation of time—all set to correspond to what may be observed in Geneva.

Vacheron Constantin

Patrimony Traditionnelle Calibre 2253 Collection Excellence Platine

The Vacheron Constantin Patrimony Traditionnelle Calibre 2253 Collection Excellence Platine is endowed with a tourbillon movement powering indications of the perpetual calendar and times of sunrise and sunset, as well as the equation of time on the arc of a circle at 10:30.

Girard-Perregaux
1945 Vintage Perpetual Calendar and Equation of Time

Housed in a rectangular Art Deco-inspired case, the 1945 Vintage Perpetual Calendar and Equation of Time from Girard-Perregaux has been released in a version with a perpetual calendar and an equation of time segment-type display that appears in the upper left corner of the dial.

Daniel Roth,
Perpetual Calendar Equation of Time

Pride of the Masters Grandes Complications collection, the Perpetual Calendar Equation of Time model from Daniel Roth boasts an open-worked dial that allows glimpses of the movement as well as the different calendar discs.

Panerai
Luminor 1950 Equation of time Tourbillon Titanio 50 mm L'Astronomo

The Luminor 1950 Equation of time Tourbillon Titanio 50 mm L'Astronomo provides an innovative linear display of the equation of time at 6 o'clock. It also indicates the times of sunrise and sunset for a reference city chosen by the client, and its caseback bears a depiction of the night sky as seen from that same city.

Taking Longitude into Account

By convention, all the places situated in one time zone have the same time. But "true" solar noon varies by longitude, and it is distinctly earlier in Warsaw than in Madrid. Executing a correction of -16 to +14 minutes over an average applicable to the entire time zone (60 minutes) is therefore not of any great scientific interest, but innovative watchmakers are remedying this flaw with models that take into account the exact degree of longitude.

< Arnold & Son, *True North Perpetual*

The True North Perpetual watch from Arnold & takes account of longitude, thanks to a dial featurin mobile rings. The model features another unprecec characteristic: when the 24-hour solar hand indicat true solar noon, the wearer need only point it towar sun for the mobile outer dial to indicate true geogra North—hence its name.

> Audemars Piguet, *Royal Oak Equation of Time*

The first brand to have developed an equation of time indication system adjusted to the exact degree of longitude, Audemars Piguet equipped its Royal Oak Equation of Time with a large central hand for the complication's display. The inner bezel ring bears the name of the reference city chosen by the client, as well as the time of true noon in that city. When the hour hand corresponds to the hour indicated on the inner bezel ring, and the equation of time and minute hands line up exactly, the owner of the watch knows that the sun is exactly at its zenith. This is also one of the rare wristwatches to display sunrise and sunset times while taking into account three parameters: date, longitude and latitude.

Running Equations of Time

Distinctly more user-friendly yet far more complex in terms of their construction, "running" equation of time watches have two hands fitted coaxially with the minute hand, respectively indicating mean time and true time. They require no mental arithmetic, since true time is continuously displayed directly on the dial. This system, which had long been regarded as the exclusive preserve of pocket-watches, first appeared on wristwatches in 2004 and represents a genuine feat in terms of miniaturization.

Jaeger-LeCoultre, *Gyrotourbillon 1*

The Gyrotourbillon 1 by Jaeger-LeCoultre, equipped with a spherical tourbillon (see chapter on Tourbillons), also features an offset hour/minute subdial bearing a running equation of time displayed by an additional minute hand adorned with a tiny golden sun.

Blancpain

Villeret Equation Marchante Pure

Blancpain has associated a running equation of time display and a perpetual calendar in a model from the Villeret collection named Equation Marchante Pure. A circular opening at 6 o'clock allows the wearer to contemplate the cam that governs the equation, as well as the sensor in action.

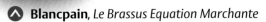

Blancpain, *Le Brassus Equation Marchante*

The Le Brassus Equation Marchante by Blancpain is distinguished by its double equation of time display: the running equation of time itself appears in the center with an additional central minute hand tipped with a golden sun; and a subdial at 2 o'clock boasts a retrograde hand indicating the number of minutes of difference. This sophisticated device is combined with an ultra-thin perpetual calendar.

AUDEMARS PIGUET

JULES AUDEMARS EQUATION OF TIME – REF. 26053PT.OO.D002CR.01

Housed in a 950 platinum case, the Jules Audemars Equation of Time features a plethora of impressive indications. The automatic-winding movement, visible through the openworked dial, powers a perpetual calendar with astronomic moon, as well as indicating the times of sunrise and sunset. Accompanying the hour and minute hands, a sinuous sun-tipped hand swings between +15 and -15 to indicate the difference between mean time and solar time.

BLANCPAIN

VILLERET EQUATION DE TEMPS MARCHANTE – REF. 6638-3631-55B

The Equation Marchante Pure, based on the Le Brassus model that was the first wristwatch capable of displaying the separate solar and civil times by two minute hands, is an integral part of the Villeret collection, characterized by its pure lines and formal beauty. Fitted with a three-subdial perpetual calendar complete with a retrograde moonphase, this red-gold masterpiece delights with an aperture at 6:00, through which the owner is treated to the precise analemma-shaped equation of time cam. This breathtaking piece houses the 364-part caliber 3863A, equipped with 72 hours of power reserve.

BREGUET

CLASSIQUE GRANDE COMPLICATION – REF. 3477BR.1E.986

Breguet has refined the Equation of Time complication with this remarkable piece. Housed in an 18K rose-gold case with a diameter of 35.5mm, this model boasts a silvered-gold dial, which is hand-engraved on a rose engine. The house's horologists have devised and patented the equation of time mechanism in this model, combined with a legible perpetual calendar. Both complications are designed to function without correction for over two centuries. The sapphire crystal caseback allows for a water resistance of 3atm. This watch is also available in yellow gold and platinum.

VACHERON CONSTANTIN

PATRIMONY TRADITIONNELLE CALIBRE 2253 – REF. 88172/000P-9495

A hand-stitched dark blue alligator strap with platinum threads traces the path to a truly astonishing spectacle. Limited to 10 numbered pieces, this 44mm 950 platinum timepiece houses a remarkable manually wound caliber honored with the Poinçon de Genève. While the perfectly finished tourbillon with small seconds at 6:00 completes its rotation in precisely one minute, the rest of the dial looks at time on a significantly grander scale. Set to the owner's reference city of choice, the sunrise and sunset indicators surround the exposed carriage on either side. On the upper half of the watch face, the perpetual calendar, complete with day of the week, month, date, and leap-year indicator, is joined by the equation of time between 10:00 and 11:00. Merging the tourbillon's symbolism of horological history with a wealth of highly sophisticated functions, the Patrimony Traditionnelle Calibre 2253 makes a magnificent statement regarding the grandeur of time.

Tourbillons

Richard Mille
RM 56-01
Tourbillon Sapphire

Richard Mille, Founder and
CEO of Richard Mille

Richard Mille

Founder and CEO of Richard Mille

"I am part of the establishment now, but I am still a revolutionary"

SINCE 2001, RICHARD MILLE HAS OFFERED (OFTEN QUITE AVANT-GARDE) WATCHES IN LIMITED PRODUCTION RUNS. In 2013, only 3,000 timepieces emerged from the Breuleux, Switzerland workshops, a number comparable to previous years, with an ever-increasing average price. Several models, such as the RM 027 Rafael Nadal or the RM 031 Haute Performance, serve as research laboratories. This year, as in the past, Richard Mille is presenting 10 new launches. An entire women's collection, with steel bracelets, is part of the lineup—a first for the brand. In the years to come, Richard Mille, Founder and CEO, predicts more "wonderful developments."

What were the main highlights of 2013 for Richard Mille?

First of all, we are quite happy with the increase in our sales numbers, which have risen to 130 million Swiss francs. The brand has also acquired a certain maturity. For example, we've smoothed out our production process, working more evenly over the course of the year. This calmer pace allows our watchmakers to devote due care to the timepieces without working overtime. Our subcontractors, such as the Vaucher and APRP (Audemars Piguet Renaud & Papi) manufactures, have also greatly contributed to this improvement in the way we organize our work.

How are your collections organized?

Richard Mille's product line always includes two segments. We produce certain models that are a permanent part of our collections, such as the RM 007, RM 009, RM 010 and RM 011. At the same time, we develop "extreme" watches, usually in collaboration with APRP. These limited editions represent the various stylistic approaches we take to horology. I love combining a few functions within one watch! In order for them to perform in concert, we really have to work hard. Once the series is finished, I move on to other innovations.

RM 027 Tourbillon Rafael Nadal

What particular expertise have you acquired over the years?

It's been ten years now since we started working on exceptional watches such as the RM 027 Tourbillon Rafael Nadal, the RM 018 Hommage à Boucheron and the RM 031 Haute Performance. Over ten years of doing this, we've mastered techniques as extreme as our watches! Certain pieces also serve as a "lab." For example, we poured into the RM 031 all the expertise gained since the brand's founding. Its mission: chronometric performance. We worked on crucial aspects, such as reducing friction. In the end, each of the 10 watches in the series boasts a precision amounting to a few seconds per month. This extraordinary performance rivals that of quartz movements. Other Richard Mille watches subsequently incorporated these techniques, leading to high scores on measures of reliability.

RM 031
Haute Performance

Will you always be so creative in your collections?

The brand was just two years old when some of my competitors started announcing the imminent end of my creativity! The truth is, every year I have to force myself to slow down. This year, we are introducing 10 new models. To my great surprise, I realized that we had never before used steel bracelets. So you see, we still have some wonderful developments in store! This year we are introducing a very strong women's collection, on a sensational metal bracelet.

Who buys Richard Mille watches?

Often, it is men who buy our women's models to give as gifts, because of the high price point. I really had to push to get our women's collections off the ground. Since 2012, though, it has exploded in popularity. I was completely out of stock throughout 2013! Women's tastes have become more sophisticated, and they are also looking for unique brands such as ours.

As far as the age of our clientele is concerned, most of our customers were around 50 years old when I started. A recent study showed, however, that Richard Mille watches are the most popular watches among the younger generations. At Watches & Wonders 2013, the crowd around our stand was largely between 35 and 40 years old. This clientele is spread all over the world— North America, South America, Europe and Asia.

How are you developing your markets in 2014?

We have the possibility for expansion in many countries, including Germany, Scandinavia, Eastern Europe and Korea, where we will be opening a boutique in Seoul. From 16 boutiques in 2013, we will increase the number to 25 by the end of 2014. We are opening in New York, Milan, Munich, Qatar, Jeddah and on London's very chic Bond Street. This spring we are moving our Paris location to a space that measures $220m^2$, where our former location only measured $17m^2$. All of these new stores are joint ventures, with the same four distribution partners: one in Japan, one for the rest of Asia, the third in the Americas and the last for Europe, the Middle East and Africa. For a niche brand like ours, there is n o substitute for a local partner. Then we wait for a good opportunity and the ideal location. We also have a good relationship with LVMH and Richemont. They usually welcome our arrival in spots where they also have retail stores, since they know we attract an excellent clientele.

RM 039 Aviation

What did you think of the first Watches & Wonders salon last year in Hong Kong?

The salon is extremely significant for us. Far from being a simple imitation of the Salon International de la Haute Horlogerie (SIHH) in Geneva, it opened its doors to the public. Clients passed through in a continuous stream. We gained new kinds of watch lovers as customers, and distributors saw record sales. Another positive effect was the encouragement of the retail sellers, who obtained access to new brands and new clients. The experience was highly enriching for them.

What are Richard Mille's goals for growth?

We would like to reach sales of 150 million Swiss francs in 2014. We will continue to emphasize a range of developments, such as our women's watches. However, production has to keep pace, so I want to grow at a comfortable rate. Every year, Richard Mille reaches the next level, and always with demand outstripping supply. In 2014, we will produce 3,300 watches, with an average price that is also on the rise.

So there's no crisis for niche brands like Richard Mille?

Demand remains enormous for our products, so we've been able to avoid a slump in sales. Our collections cover several fields: technical feats, sports, lifestyle and women's watches. This assembly of niche products makes the brand universal, complete and coherent. Despite a small production volume, Richard Mille is now an essential, established brand. People think I may turn bourgeois, but even though I'm part of the establishment now, I am still a revolutionary!

What ambassadors and events will you be involved with in 2014?

We are going in a more cultural direction, with artistic, musical and dance events. I have always said that we needed to get out of the sports ghetto and get involved with the art world. We were also partners in the Takashi Murakami show at Art Basel Hong Kong in 2013. We remain partnered with Rafa (Rafael Nadal) on very long-term contracts. Felipe Massa also remains one of our close partners. At the same time, Richard Mille is continuing as the official timekeeper for the Lotus F1 team.

⌃ RM 051 Phoenix – Michelle Yeoh

Tourbillons

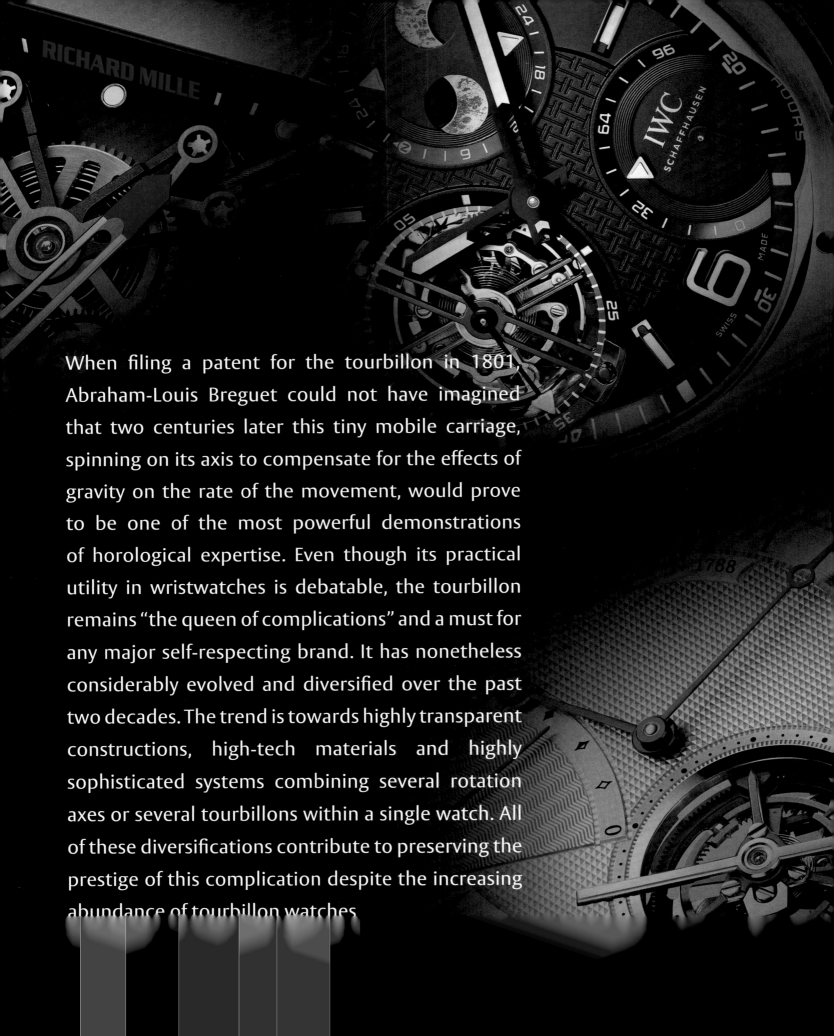

When filing a patent for the tourbillon in 1801, Abraham-Louis Breguet could not have imagined that two centuries later this tiny mobile carriage, spinning on its axis to compensate for the effects of gravity on the rate of the movement, would prove to be one of the most powerful demonstrations of horological expertise. Even though its practical utility in wristwatches is debatable, the tourbillon remains "the queen of complications" and a must for any major self-respecting brand. It has nonetheless considerably evolved and diversified over the past two decades. The trend is towards highly transparent constructions, high-tech materials and highly sophisticated systems combining several rotation axes or several tourbillons within a single watch. All of these diversifications contribute to preserving the prestige of this complication despite the increasing abundance of tourbillon watches.

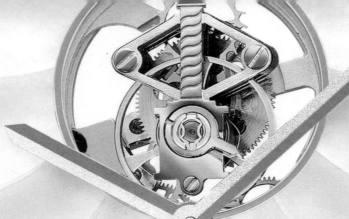

THE TOURBILLON WAS CREATED TO COMPENSATE FOR FLUCTUATIONS IN A POCKET-WATCHES FUNCTIONING CAUSED BY THE EARTH'S GRAVITATIONAL PULL.

Maintained in a vertical position for many hours, the pocket-watch mechanism—starting with the balance and the balance-spring, two necessary components for accuracy and precision—underwent the detrimental effects of gravity. To avoid such irregularities, the famous watchmaker Abraham-Louis Breguet (1747-1823)—one of the greatest horological geniuses of all time—had the idea of enclosing the "heart" of the watch (balance, balance-spring and escapement) within a small mobile carriage spinning on its axis within a predetermined time (generally one, four or six minutes). It was for this rotation that he chose to name his invention "tourbillon"—a term from astronomy synonymous with "planetary system" among 18th-century scientists. Breguet submitted the patent for his "watch compensating for all the inequalities that occur in the balance and balance-spring" on the 7th of Messidor in the 9th year of the Republican calendar—otherwise known as June 6, 1801. According to the house's archives, 35 tourbillon watches were sold between 1805 and 1823, and throughout the 19th century the tourbillon remained an extremely rare complication, reserved for a few exceptional timepieces.

The 20th-century triumph of the wristwatch could have brought about the disappearance of the tourbillon, as the complication loses much of its significance when the watch moves in all directions with the movement of the wearer's wrist. However, the revival of mechanical horology in the 1980s propelled this mechanism into the firmament of horological complications by positing it as the nec plus ultra of Swiss savoir-faire, conferring upon any watchmaker entry into the ranks of the great. Since the turn of the millennium, the tourbillon has become wildly popular. Whereas only 670 tourbillons were produced between 1801 and 1986, today the global production stands at well over 1,000 per year. Over 50 Swiss brands offer them among their collections, which is surprising when one considers that only a small handful of artisans are really capable of making this small magic carriage... Faced with this (relative) banalization, the major watch firms are determined to offer "in-house" tourbillons featuring new technical and aesthetic developments.

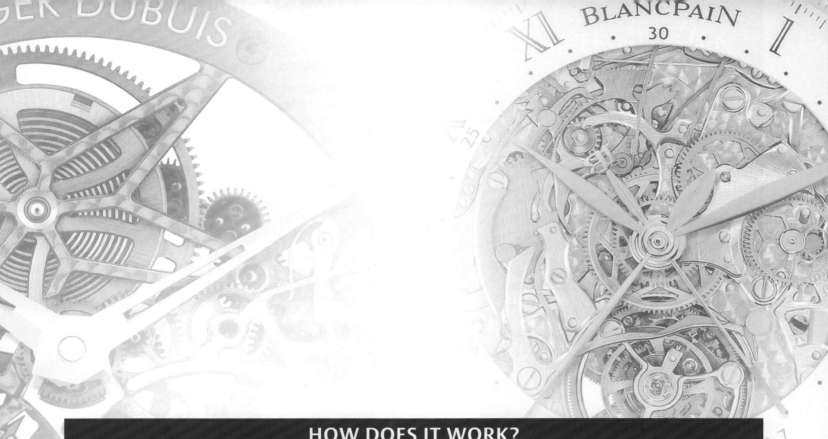

HOW DOES IT WORK?

Placed horizontally, a watch's regulating organ (balance, balance-spring and escapement) is uniformly affected by the Earth's gravitational pull. However, when it is placed vertically (and to a lesser extent, in any position besides the horizontal), gravitational force causes certain fluctuations in its functioning and oscillations. These small accelerations and decelerations cause irregularities in the watch's functioning. The idea behind the tourbillon is to place the regulating organ in a small mobile carriage that completes one rotation in a given time, usually one minute. By successively adopting all possible positions, the carriage allows the small irregularities to cancel each other out. The tourbillon carriage is a tiny construction that may contain 70 components with a weight of less than 1g. This is why the production, assembly and setting of a tourbillon watch require exceptional expertise, reserved for the most elite watchmakers and incompatible with mass production.

TOURBILLONS AND PRECISION

The tourbillon was created for the pocket watch, which spends most of its time in a vertical position. It is less necessary on a wristwatch, when the daily movements of the wearer's wrist play the same role as the tourbillon in putting the regulating organ in a variety of positions. Some people go so far as to say that the tourbillon is completely useless in its classic form. To solve this problem, a new generation of tourbillons has arrived—with diverse variations on the manners, axes and speeds of rotation—that supposedly offers enhanced performances for wristwatches. Nonetheless, even though the tourbillon is not in itself synonymous with increased accuracy, the expertise necessary to create such a watch, and the extreme care involved in its production, are both guarantees of a watch that offers superior performance.

FLYING TOURBILLON

Normal tourbillons are supported by two bridges, one above the mechanism and the other below. "Flying" tourbillons are supported by only one, below the mechanism, which allows the watchmaker to give the construction a more ethereal and resolutely virtuoso appearance.

TOURBILLON AND KARUSSEL

Patented by the Danish Bonniksen in 1892, the karussel has the same ultimate goal as the tourbillon. However, it is distinguished (or could be distinguished—the experts do not all agree) by the double wheel train that links it to the barrel: the first provides the necessary energy for the functioning of the escapement, and the second controls the speed of the carriage's rotation.

A Reinvented Classic

With its mobile carriage spinning on its axis like a planet, the tourbillon provides a fascinating sight that most brands are keen to exhibit through the dial, traditionally through an aperture at 6 o'clock. But the ultra-sophisticated mechanism is now the object of numerous technical and aesthetic innovations, such as the use of new materials that offer enhanced lightness, or the search for unusual dial configurations.

Patek Philippe, *10-Day Tourbillon, Ref. 5101*

Endowed with a power reserve of ten days, the 10-Day Tourbillon, Ref. 5101 from Patek Philippe is notable for its cambered Art Deco-style case. As with all its tourbillon watches, the manufacture prefers not to show the tourbillon from the dial side, for fear that sunlight might dry up the lubricating oils.

IWC

Ingenieur Constant-Force Tourbillon

As its name implies, the Ingenieur Constant-Force Tourbillon by IWC has a constant-force mechanism guaranteeing virtually invariable amplitude of the balance and thus enhanced precision.

Montblanc, *ExoTourbillon Chronographe*

On its ExoTourbillon Chronographe, from the Villeret 1858 collection, Montblanc has placed the balance on the exterior of the tourbillon carriage. As the balance is larger, its amplitudes gain more precision, and the tourbillon carriage, freed from the weight of the balance, requires less energy for its rotation.

Parmigiani, *Pershing Tourbillon Abyss*

A combination of titanium and pink gold, the Pershing Tourbillon Abyss by Parmigiani displays its 30-second tourbillon on a dark blue dial adorned with a wave motif—a nod to the famous Italian yacht maker after which the collection is named.

A. Lange & Söhne

Richard Lange Tourbillon "Pour le Mérite"

Beneath the regulator-type dial of the Richard Lange Tourbillon "Pour le Mérite" watch lie two mechanisms initially designed to improve timekeeping precision: a tourbillon and a fusée-and-chain energy transmission system.

F.P. Journe, *Tourbillon Historique*

To celebrate his 30 years of horological creativity, watchmaker F.P. Journe has launched a Tourbillon Historique wristwatch strongly inspired by his first pocket-watch. The tourbillon is visible through a sapphire caseback concealed beneath a hinged cover.

> **Breguet**, *Classique* 5377

A worthy heir to Abraham-Louis Breguet's famous patent, the Classique 5377 combines a silicon balance-spring and lever-wheel, a titanium balance-wheel and tourbillon carriage, and a platinum peripheral rotor—all housed within an extraordinarily slim (3mm) movement.

Richard Mille, *Tourbillon RM 053 Pablo Mac Donough*

Designed to withstand the extreme shocks encountered during a polo match, Richard Mille's Tourbillon RM 053 Pablo Mac Donough (dedicated to one of the sport's most eminent stars) is equipped with two "windows," the left one for the tourbillon and seconds display, the right one for the hours and minutes.

Vacheron Constantin
Patrimony Traditionnelle 14-Day Tourbillon

Using four barrels in paired series, the Patrimony Traditionnelle 14-Day Tourbillon displays a power reserve of 336 hours. The tourbillon's carriage is adorned with the Maltese Cross, Vacheron Constantin's emblem.

Chopard, *L.U.C Engine One H*

In addition to the central hour and minute hands sweeping over a horizontal dial reminiscent of racing car instrument panels, the L.U.C Engine One H by Chopard also provides a power reserve display on the left and a tourbillon topped with a small seconds hand on the right.

Antoine Preziuso
GTS Grand Tourbillon Sport

Equipped with a cobalt chrome case, whose aerodynamic forms evoke the sportscars that inspired them, Antoine Preziuso's GTS Grand Tourbillon Sport is also notable for its large red hands, which recall the needles on speedometers.

Room for Transparency

Certain brands have chosen to distill the tourbillon's magic with constructions whose heart is open, revealing the entire interior of the movement—masterpieces of skeletonization that push the frontiers of possibilities, or fascinating transparent stage-settings within which the mobile carriage seems to float weightlessly.

> **Corum**
> *Golden Bridge Tourbillon*
>
> Corum shows off its famous "baguette" movement in this Golden Bridge Tourbillon model, with a sapphire bridge. In order to reveal the flying tourbillon, a part of the movement (usually in gold) has been replaced by a sapphire component.

∧ **RJ-Romain Jerome**, *Moon Orbiter*

Issued in a 25-piece limited edition, the Moon Orbiter by RJ-Romain Jerome reveals a flying tourbillon on the left side of the dial. The steel case incorporates fragments from the Apollo 11 capsule, while the black dial contains genuine moon dust.

> **Montblanc**, *Tourbillon Bi-Cylindrique*

The Tourbillon Bi-Cylindrique from Montblanc stands out for its architecture with openworked heart, double cylindrical balance-spring and its "mysterious hours" display that seems to float in the air thanks to the use of crystal discs.

> **Haldimann**, *H8 Sculptura*

On its H8 Sculptura, Haldimann has eliminated the hour and minute indications, retaining only the fascinating spectacle of the central tourbillon floating in between two sapphire crystals. It is a true work of art, constructed entirely with traditional techniques.

∧ **Richard Mille**, *RM 27-01 Rafael Nadal*

Designed in cooperation with the international tennis star, the RM 27-01 Rafael Nadal by Richard Mille combines a feather-light weight (3.5g for the movement) with a high-tech construction based on a carbon case and a cable-suspension mechanism.

De Bethune, *DB 28T*

The DB 28T from De Bethune stands out for its futuristic architecture and its ultralight tourbillon in titanium and silicon. The tourbillon carriage rotates over 30 seconds and the balance oscillates at 36,000 vph.

Revelation

R04 Tourbillon
Magical Watch Dial®

The R04 Tourbillon Magical Watch Dial® by Revelation comprises a patented system serving to make the movement appear as if by magic through a dial that has turned transparent. It is also equipped with a Tourbillon Manège® caliber featuring a rotating regulating system in which the regulating organ is mounted on a mobile arm sandwiched between two wheels mounted on the central axis.

Piaget

Emperador Coussin Tourbillon
Automatic Ultra-Thin

Equipped with an exceptionally thin manufacture movement, the Emperador Coussin Tourbillon Automatic Ultra-Thin from Piaget stands out for its unusual geometry, with an off-center hours and minutes display, tourbillon carriage at 1:30 and a micro-rotor visible on the dial at 9 o'clock.

Girard-Perregaux
Laureato Tourbillon with three bridges

In 1889, Girard-Perregaux made watch-making history by creating its famous "tourbillon with three gold bridges." In its Laureato Tourbillon with three bridges, the manufacture revisits this legacy by associating a titanium case with three translucent bridges in blue spinel.

Audemars Piguet
Jules Audemars Large Date Tourbillon

The Jules Audemars Large Date Tourbillon from Audemars Piguet, powered by a manual-winding in-house caliber, opts for understated elegance with a large tourbillon carriage and a panoramic date display at 12 o'clock.

Blancpain
Le Brassus Tourbillon Carrousel

Blancpain has innovated by rehabilitating the carrousel (or karussel), a mechanism similar to the tourbillon (see "How does it work?" section) and incorporating it within several wristwatches, including the Le Brassus Tourbillon Carrousel with flying tourbillon at 12 o'clock and flying carrousel at 6 o'clock.

The Third Dimension

Recent years have also witnessed the appearance of a new generation of tourbillons equipped with inclined carriages or several simultaneous rotational axes, and/or revolving at speeds that diverge from the traditional one minute (often 30 or 24 seconds). Besides their often spectacular appearance, these three-dimensional constructions are designed to be more suited to wristwatches, by multiplying the positions of the balance-spring with the case.

∧ **Greubel Forsey**, *Art Piece 1*

On the Art Piece 1 by Greubel Forsey, the Double Tourbillon 30° (with its inner carriage rotating in one minute and the outer carriage in four minutes), vies for attention with a microsculpture created by contemporary artist Willard Wigan.

∧ **Concord**, *Quantum C1 Gravity*

Concord created shock waves with its launch of the Quantum C1 Gravity, equipped with an unprecedented bi-axial tourbillon positioned vertically, i.e. almost outside the movement, and linked to the bottom plate by a series of cables reminiscent of suspension bridges.

> **Jaeger-LeCoultre**, *Duomètre Sphérotourbillon*

Equipped with a Dual-Wing movement, Jaeger-LeCoultre's Duomètre Sphérotourbillon can rightfully claim to be the first tourbillon watch adjustable to the second. The "sphérotourbillon" rotates on two axes: one rotation is completed in 30 seconds, and the other (on a 20° angle) in 15 seconds.

Harry Winston, *Histoire de Tourbillon 4*

The Histoire de Tourbillon 4 model by Harry Winston is equipped with a tri-axial flying tourbillon starring beneath a domed sapphire crystal, featuring an inner carriage rotating in 45 seconds, an intermediate carriage in 75 seconds and an outer carriage in 5 minutes.

Panerai, *Radiomir Tourbillon GMT Ceramica Lo Scienzato*

On the Radiomir Tourbillon GMT Ceramica Lo Scienzato from Panerai, the carriage turns at a right angle to the oscillation axis of the balance, performing two turns per minute. Its in-house Caliber P.2005/S is also distinguished by a substantial six-day power reserve, achieved thanks to the energy from three barrels.

Double Revolutions

A few innovative models offer a sophisticated system in which the tourbillon completes a rotation in one minute, while performing a complete turn of the dial in one hour (or more)—mirroring the double rotation of the Earth on its axis and around the sun. These are truly magical merry-go-rounds.

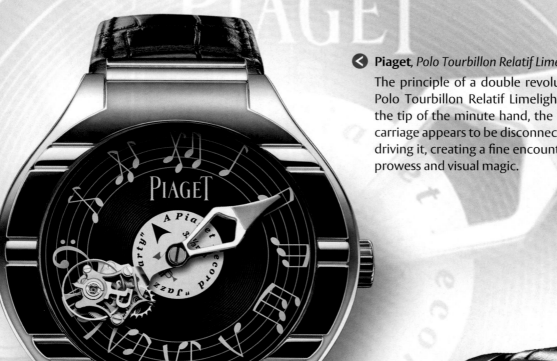

Piaget, *Polo Tourbillon Relatif Limelight Jazz*

The principle of a double revolution applies to Piaget's Polo Tourbillon Relatif Limelight Jazz. Suspended from the tip of the minute hand, the mobile flying tourbillon carriage appears to be disconnected from the movement driving it, creating a fine encounter between mechanical prowess and visual magic.

Cartier
Rotonde de Cartier Astrotourbillon
Carbon Crystal

The Rotonde de Cartier Astrotourbillon Carbon Crystal has an original construction with a one-minute central tourbillon fitted above the movement. The arrow-shaped balance bridge indicates the seconds. The case is com-posed of a niobium-titanium alloy, while the tourbillon bridges, lever and the escape-wheel are in carbon crystal.

Ulysse Nardin, *Freak Phantom*

The Freak Phantom from Ulysse Nardin stands out for its very innovative construction. The movement rotating within the case allows the watch to display the hours and minutes with neither hands nor dial. The flying karussel tourbillon indicates the seconds. The escapement and balance spring are in silicon.

Cartier, *Rotonde de Cartier Mysterious Double Tourbillon*

The Rotonde de Cartier Mysterious Double Tourbillon is distinguished by its flying tourbillon featuring a double rotation (once on its own axis in 60 seconds, and around an aperture every 5 minutes) and of which the carriage appears to be floating in mid-air thanks to a transparent disc-based system.

Bernhard Lederer, *Gagarin Tourbillon*

A tribute to the famous Soviet cosmonaut, the Gagarin Tourbillon by Bernhard Lederer displays an orbital 60-second tourbillon also rotating counter-clockwise in 108 minutes. A prominent loupe, held in place by a catch inspired by a hatch on the Vostok space capsule, affords a chance to admire the dial details.

Watches with Several Tourbillons

Another technical and aesthetic path explored by some watchmakers is the multiplication of tourbillons within a single watch. Horologers come up with double, even quadruple tourbillons, tied together by differential gear systems that average out the functioning of different regulating organs.

∧ **TAG Heuer**, *Carrera Mikrotourbillons*

The Carrera Mikrotourbillons from TAG Heuer possesses two tourbillons, one for the time, the other for the 1/100-of-a-second chronograph. The latter, described as "the fastest tourbillon in the world," beats at 360,000 vph and completes 12 revolutions per minute.

< **Roger Dubuis**
Excalibur Double Tourbillon Squelette

Boasting two flying tourbillons linked by a differential, the Excalibur Double Tourbillon Squelette model reveals all the technical sophistication and aesthetic refinement of its 292-component movement, graced by the prestigious Poinçon de Genève.

Greubel Forsey, *Invention Piece 2*

A reinterpretation of its 2008 Quadruple tourbillon à différentiel, Greubel Forsey's Invention Piece 2 possesses two double tourbillon carriages (a first carriage executing a complete rotation in one minute, contained within a second carriage effecting a complete revolution in four minutes), linked by a spherical differential placed at the center of the timepiece.

Jacob & Co., *Napoleon Quadra Tourbillon*

Crafted in titanium, the imposing Napoleon Quadra Tourbillon from Jacob & Co. features no fewer than four skeletonized tourbillons and four complete time zones with hour and minute displays.

Breguet
Twin Rotating Tourbillons

The Twin Rotating Tourbillons from Breguet boasts two independent tourbillons coupled by a differential gear and fixed to a bottom plate completing a full rotation in 12 hours. These two mechanisms mutually compensate for their infinitely small variations in rate in order to offer superior precision.

The Feminine Tourbillon

Within just a few years, women's tourbillon watches have accomplished an astonishing breakthrough. Uniting haute horology and—most often—haute jewelry, the most prominent brands have undertaken to create timepieces that set a refined stage for the little "magic carriage."

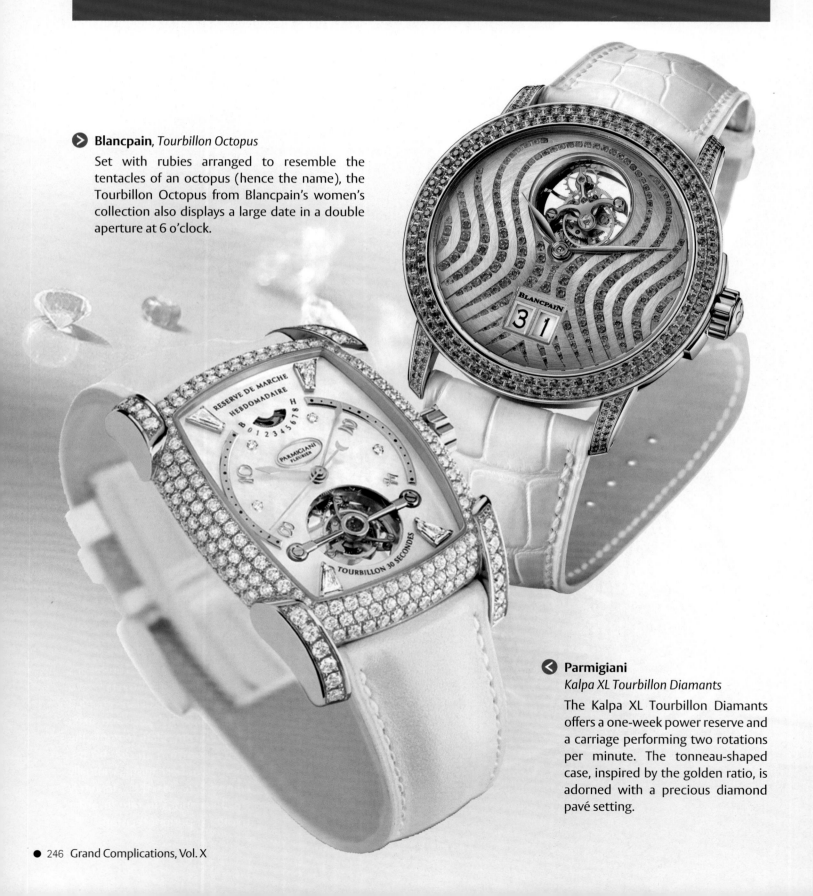

Blancpain, *Tourbillon Octopus*

Set with rubies arranged to resemble the tentacles of an octopus (hence the name), the Tourbillon Octopus from Blancpain's women's collection also displays a large date in a double aperture at 6 o'clock.

Parmigiani

Kalpa XL Tourbillon Diamants

The Kalpa XL Tourbillon Diamants offers a one-week power reserve and a carriage performing two rotations per minute. The tonneau-shaped case, inspired by the golden ratio, is adorned with a precious diamond pavé setting.

Ulysse Nardin, *Royal Ruby Tourbillon*

Adorned in rubies and diamonds, the Royal Ruby Tourbillon by Ulysse Nardin sets a fine stage for its flying tourbillon within a supremely transparent structure reflecting the red glow of the precious stones. This platinum limited edition is available in 99 pieces.

◄ **Louis Vuitton**
Tambour Monogram Tourbillon

The Tambour Monogram Tourbillon by Louis Vuitton plays on the House signature codes, including the famous LV initials. The tourbillon carriage, visible at 6 o'clock, is equipped with a bridge shaped like the flower on the famous Monogram canvas.

► **Van Cleef & Arpels**
Midnight Tourbillon Nacre

The superbly understated Midnight Tourbillon Nacre from Van Cleef & Arpels is available in a limited edition of 100 watches. The handcrafted mother-of-pearl dial reveals a tourbillon carriage that echoes the style of the Vendôme Column in Paris.

A. LANGE & SÖHNE

1815 TOURBILLON – REF. 730.025

A micro-mechanical picture of supreme precision, this hand-wound platinum timepiece complements its exhibited tourbillon with a sophisticated duo of patented mechanisms. A pull of the 3:00 crown instantaneously halts the tourbillon cage's balance thanks to the stop-seconds function, while the L102.1 caliber's zero-reset mechanism drives a simultaneous jump of the seconds hand to the zero position, permitting an exact synchronization with a chosen time reference. Animated by two blued steel hands for hours and minutes on a solid silver dial, the limited edition 1815 Tourbillon is worn on a hand-stitched crocodile strap.

A. LANGE & SÖHNE

LANGE 1 TOURBILLON PERPETUAL CALENDAR – REF. 720.025

The Lange 1 Tourbillon Perpetual Calendar's self-winding L082.1 caliber is ingeniously fitted with a platinum centrifugal mass for exceptional winding efficiency. Thanks to an innovative rotating peripheral month ring that occupies a minimal amount of space, and an hour and minute circle off-centered on the solid silver rhodiumed dial, A. Lange & Söhne accomplishes a perpetual calendar of remarkable ease of use and legibility, despite its comprehensive constitution of moonphase, retrograde day, leap-year aperture and date. Admired intimately through the sapphire crystal caseback of this 41.9mm platinum masterpiece, the exquisite tourbillon is embellished with an extremely sophisticated black polished finish, and boasts a flawless diamond end stone as its radiant bearing.

A. LANGE & SÖHNE

RICHARD LANGE TOURBILLON "POUR LE MERITE" – REF. 760.032

The Richard Lange Tourbillon "Pour le Mérite" is powered by the manual-winding new hand-finished Lange manufacture L072.1 caliber. It incorporates a fusee-and-chain transmission as well as a tourbillon with a patented stop-seconds mechanism and a balance spring manufactured in-house. The 41.9mm case comes in platinum or rose gold with a 36-hour power reserve. Its dial features three intersecting circles for the time indicators. The largest dial indicates the minutes while the smaller subdials represent the seconds and hours. The watch includes a crocodile strap secured with a solid gold buckle. The Richard Lange Tourbillon "Pour le Mérite" in platinum is limited to 100 pieces.

AUDEMARS PIGUET

JULES AUDEMARS DRAGON TOURBILLON – REF. 26569OR.OO.D088CR.02

The Jules Audemars Dragon Tourbillon is powered by Audemars Piguet's in-house, hand-wound Manufacture caliber 2906, which is comprised of 215 parts and beats at 21,600 vph. This timepiece has an 18K pink-gold case and pink-gold enamel dial that features an engraved 18K pink-gold dragon. The strap is hand-stitched, large square scale crocodile with an 18K gold AP folding clasp. It is water resistant to 20m. There is a minimum guaranteed power reserve of 72 hours. This model is available in a limited edition of 18 pieces.

AUDEMARS PIGUET

JULES AUDEMARS LARGE DATE TOURBILLON – REF. 26559OR.OO.D002CR.01

Home to Audemars Piguet's 228-part 2909 caliber, this masterpiece showcases, via a stunning aperture at 6:00, a tourbillon that unarguably merits its prominent exhibition on the dial. The exposed heartbeat of the watch is balanced by the presence of an innovative large date at 12:00. The ingenious use of two superimposed dials, working in precise synchronization and without a separation of the tens and units, is crucial in permitting a date display of such magnitude. The Jules Audemars Large Date Tourbillon boasts a 72-hour power reserve, beats at a frequency of 21,600 vph, and offers the choice of black or silver-toned dial, each of which is limited to 25 pieces.

AUDEMARS PIGUET

ROYAL OAK EXTRA-THIN TOURBILLON – REF. 26510OR.OO.1220OR.01

The Royal Oak Extra-Thin Tourbillon is powered by the hand-wound, in-house Manufacture caliber 2924, enclosed in an 18K pink-gold 41mm case. The movement is comprised of 216 parts and beats at 21,600 vph. It features a blue dial with the "Petite Tapisserie" pattern and pink-gold applied hour markers. This piece includes the following functions: tourbillon, hours, minutes and power reserve. The bracelet is also made of 18K pink gold with an AP folding clasp. It is water resistant to 50m and has a minimum power reserve of approximately 70 hours.

BELL & ROSS

BR MINUTEUR TOURBILLON PINK GOLD

This tourbillon is powered by a manual-winding mechanical movement and its bridges are in pink gold and anodized aluminum. The watch displays the hours, minutes, small seconds, a three-day power reserve indicator and a flyback function (2 counters: 60 minutes and 10 tenths of an hour), which features a zero-setting and quick reset. The case is in 18K pink gold and measures 44x50mm. The black dial possesses hands and indexes treated with a photoluminescent coating for nighttime reading. Water resistant to 50m, the watch is available on a rubber or alligator strap. Limited edition of 30 pieces.

BELL & ROSS

BR01 TOURBILLON CARBON FIBER ROSE GOLD

This limited edition model represents quality watchmaking at its finest. The BR01 Tourbillon is housed in a stunning satin-polished pink-gold case with an XXL diameter of 46mm. The tourbillon cage rests on the dial at 6:00 alongside an hour counter at 12:00, a central minutes hand, a power reserve indicator at 9:00 and lastly a trust index at 3:00. The hands and indexes on the black carbon fiber dial are coated in a photoluminescent material, guaranteeing optimal legibility for the wearer. The BR01 Tourbillon is also available with titanium DLC finishing on the case or with an aluminum fiber dial.

BELL & ROSS

BR01 TOURBILLON AIRBORNE

This limited edition of 20 pieces has a manual-winding mechanical movement and contains carbon-fiber bridges and plates and a black-gold tourbillon carriage. Hours can be read in a counter at 12:00, a trust index is positioned at 3:00, and the 120-hour power reserve indicator is at 9:00. A central hand displays the minutes. The three-dimensional dial has a black mesh-like design and is treated with a photoluminescent coating for nighttime reading. The 46x46mm case is titanium with a black DLC (Diamond-Like Carbon) coating. Water resistant to 100m, the watch features a shagreen strap.

BELL & ROSS

WW2 MILITARY TOURBILLON

Bell & Ross pays homage to horology during the First World War with this exceptional timepiece. The 45mm titanium case is equipped with a protective lid, which served as armor plating during the war. This lid is now repurposed as a frame to display the complications and mechanisms on the dial. When opened via the integrated hinge, one can see the entirety of the matte black dial complete with a marvelous pink-gold tourbillon carriage at 6:00 and Côtes de Genève decor. This special model is part of an exclusive limited edition of only 20 pieces.

BLANCPAIN

TOURBILLON CARROUSEL – REF. 2322-3631-55B

Housed in an 18K red-gold case with a diameter of 44.6mm, the Tourbillon Carrousel exemplifies the strong watchmaking tradition at Blancpain. This model is equipped with a tourbillon at 12:00 and a carrousel at 6:00, along with a hand-guided date display at 3:00. A power reserve indicator can be found on the transparent caseback. Gold Roman numeral hour markers emerge from the Grand Feu enamel chapter ring. This model is powered by a 2322 caliber with a power reserve of 168 hours.

BREGUET

CLASSIQUE TOURBILLON EXTRA FLAT – REF. 5377BR.12.9WU

The Classique Tourbillon Extra Flat is distinguished by its off-centered tourbillon, which is protected by several patents. The tourbillon carriage and the balance are in titanium, the balance-spring is made of silicon and the escapement is in silicon and anti-magnetic steel. The new caliber 581DR fitted into this timepiece beats at 4Hz — a high frequency for a tourbillon — without sacrificing power reserve. The patented "high-energy" barrel provides 90 hours of running time. To keep the height of the movement down to 3mm in a 7mm-thick case, Breguet's watchmakers have placed the bi-directional platinum winding rotor on the edge of the movement.

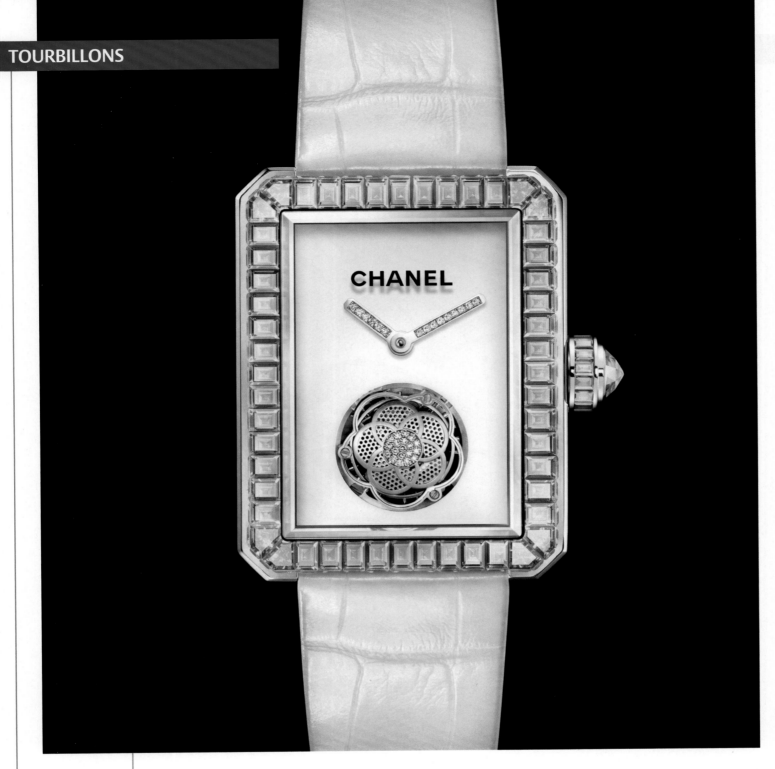

CHANEL

PREMIERE FLYING TOURBILLON – REF. H3448

CHANEL makes its first foray into the world of feminine mechanical complication watches with its Première Flying Tourbillon, which owes as much to its purity and poetry as to its complexity and virtuosity of construction. Almost secretly, the Flying Tourbillon beats within its stylized case in the form of a camellia, Mademoiselle Chanel's favorite flower. The absence of an upper bridge magnifies the Tourbillon's aesthetic—its frame becomes a micro-mechanical marvel in a three-dimensional assemblage of thin elements. This exceptional Tourbillon was developed in collaboration with Renaud & Papi Manufacturing, the high-end research and development wing of Audemars Piguet. Available in limited edition of 5 pieces in 18K white gold, diamonds and pink sapphires, the Première Flying Tourbillon combines haute horlogerie and high jewelry.

DE BETHUNE

DB28T BLACK GOLD TOURBILLON

The beauty of the case on the DB28T Black Gold is unparalleled. It is distinctly shaped and crafted in shining rose gold at a diameter of 42.6mm. The prominent lugs are composed of anthracite zirconium, which contrast with the brilliant hue of the bezel. The dark masculine tone of the lugs matches both the extra-supple alligator leather strap and the futuristic looking dial. The tourbillon carriage composed of silicon and titanium is at 6:00.

DE BETHUNE

DB28T BLACK TOURBILLON

With a power reserve of five days, the black version of DB28T houses a manual-wound movement complete with 35 jewels. Beating at 36,000 vph, this hand-crafted movement is finished with mirror-polished steel Côtes De Bethune décor. The magic of this piece is the ultra-light De Bethune silicon and titanium tourbillon, which sits prominently on the dial at 6:00. This complication perfectly complements the black-mirror-polished and silver-toned minutes ring and black-mirror-polished stainless steel bridge. The sleek anthracite zirconium case also evinces fine craftsmanship.

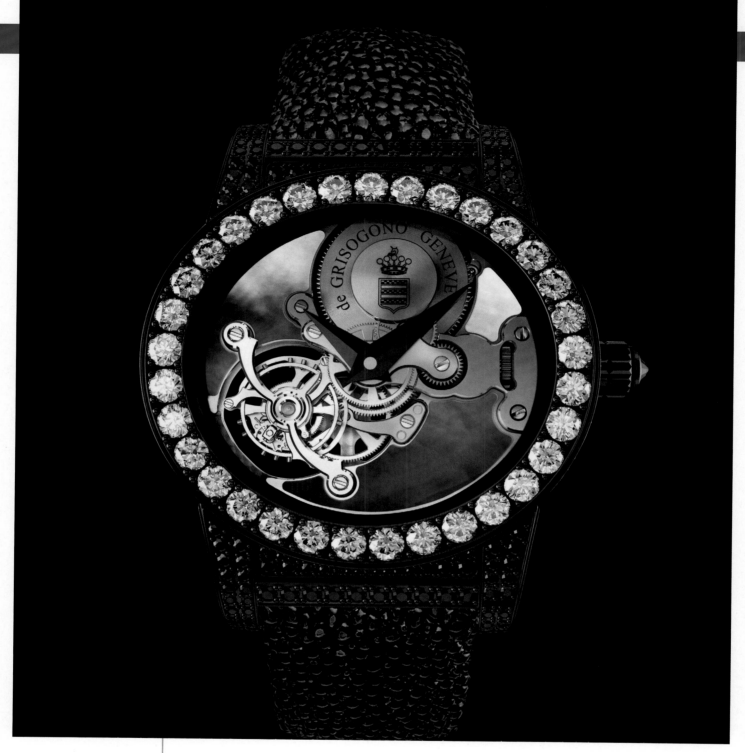

de GRISOGONO

TONDO TOURBILLON GIOIELLO – REF. TONDO TOURBILLON GIOIELLO S02

Framed by a case in black PVD-coated white gold set with 496 black diamonds, and illuminated by 33 white diamonds set onto the bezel, this hand-wound wristwatch's black mother-of-pearl dial is home to a marriage of artistic delicacy and technical sophistication. The timepiece's intricate tourbillon is revealed through an opening at 8:00, while the movement's finely finished components lead the eye to two dauphine hands indicating the hours and minutes.

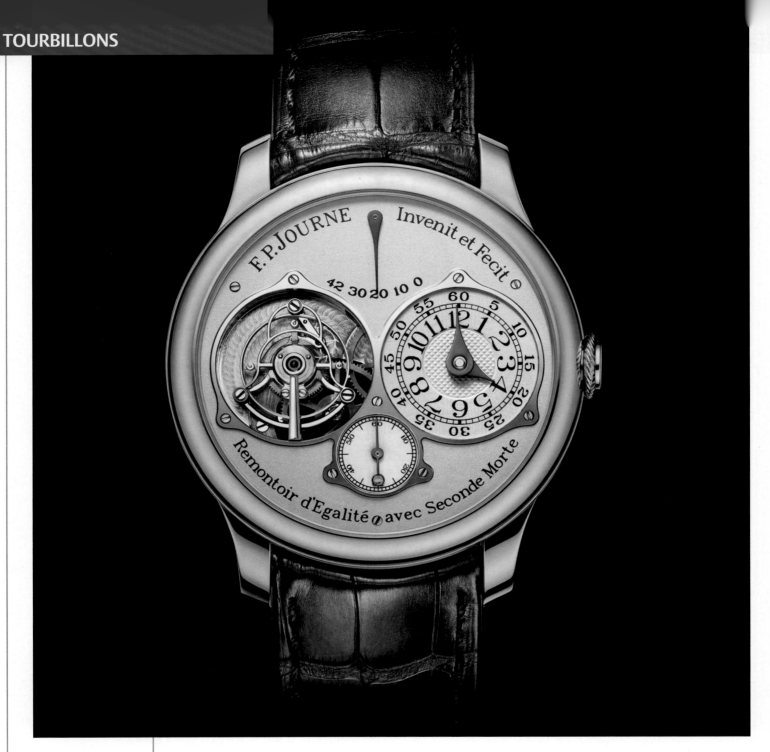

F.P. JOURNE

TOURBILLON SOUVERAIN

The new Tourbillon Souverain possesses the aesthetic characteristics and technical demands of all F.P. Journe creations. Its mechanism uses the remontoir or constant-force device system, and has now been enriched with an independent seconds system, which provides a more accurate time read-off, as the seconds hand does not move until the second has fully elapsed. This exceptional mechanism is housed in an equally noble 38 or 40mm case in 18K white or red gold, and the technical and aesthetic feat is revealed through the transparent sapphire crystal caseback. The face features the distinctive identity of F.P. Journe chronometers, with an 18K red- or white-gold dial and guilloché silver subdials screwed to the watch face (a patented feature). The 42-hour power reserve indication at 12:00 perfectly balances the independent seconds display at 6:00.

FRANCK MULLER

GIGA TOURBILLON MATRIX

This remarkable timepiece is perfectly named as it houses one of the largest tourbillons in the world. The Giga Tourbillon Matrix showcases the brilliance and audacity of Franck Muller's research and development team. Its Ø 20mm tourbillon, which takes up about one half of the watch, is powered by four Ø 16mm barrels. These extra-wide barrels are paired in parallel series that double the power of the operating reserve. This results in a continuous and reliable force. This movement has been reversed, with the bridges placed on the dial side, making the watch a rare gem. The incomparable tourbillon is housed in an equally unparalelled and stunning cage. The 18K red-gold case and cage are completely studded with over 10 carats of diamonds, making this watch perfect for a watch enthusiast with pizzazz and a stylish flair.

FRANCK MULLER

THUNDERBOLT TOURBILLON

The Thunderbolt Tourbillon, also known as the Tourbillon Rapide, is the world's quickest tourbillon mechanism. The tourbillon cage, which is powered by four barrels, makes one full rotation every five seconds. In other words, the complication rotates at a whopping 12 times per minute! The rotating frame is distinguished by a patented FM escapement with an escapement wheel, balance wheel, reversed anchor and of course, Breguet hairspring. Entirely manufactured in house, this speedy model's movement beats at 21,600 vph with a power reserve of 60 hours. It is also engraved, rhodium-plated, grained and hand beveled. This intricate movement is visible through a skeletonized dial. Its strong, sleek case is the perfect complement to this horological masterpiece.

FREDERIQUE CONSTANT

SLIMLINE TOURBILLON MANUFACTURE – REF. FC-980C4SZ9

The Slimline Tourbillon Manufacture is powered by the FC-980 caliber, which has been completely developed in-house at Frédérique Constant. The technological genius behind this movement, which in itself is comprised of over 200 components, is further enhanced by a silicium escapement wheel and anchor, providing reliability and accuracy. The gleaming hand-polished rose-gold case is an impressive 43mm in diameter and a mere 0.12cm in thickness, hence the slimline name. The dial is protected by a convex sapphire crystal, which provides the wearer with an incredible view of the Tourbillon's beating heart at 6:00.

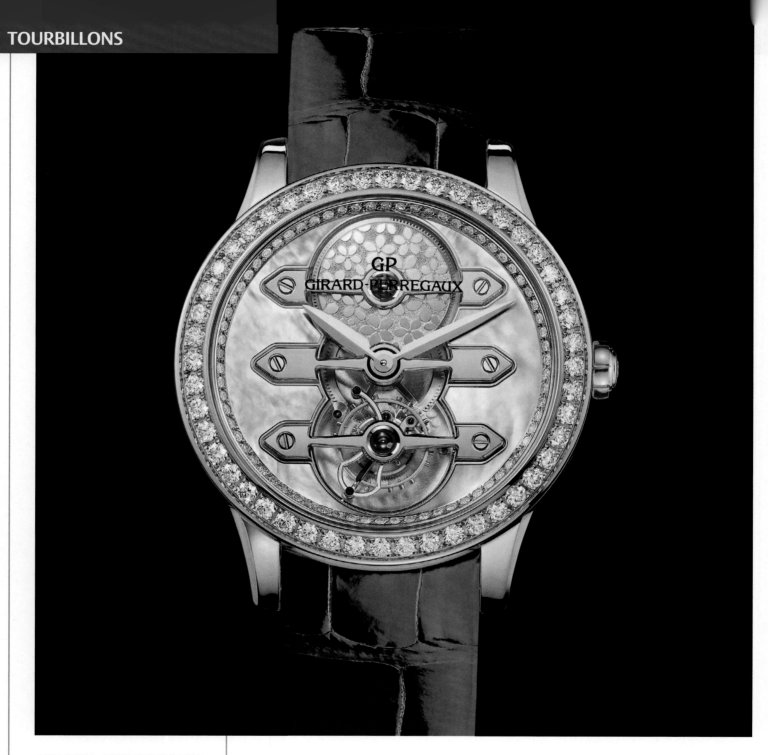

GIRARD-PERREGAUX

GIRARD-PERREGAUX TOURBILLON WITH THREE GOLD BRIDGES, LADY
REF. 99240D-52-A701-CK7A

The renowned Girard-Perregaux Tourbillon with three gold bridges enjoys a new interpretation in this feminine version. Crafted in pink gold, the subtly profiled case gracefully fits the most delicate wrists. The bezel and flange are set with more than 1.80 carats of diamonds, illuminating the tourbillon and its bridges. Their perfect geometry is underscored by entirely hand-polished finishing, while the gentle, rounded proportions seem to be magnified by an exceptional light-reflecting mirror polish. This Lady 38mm version sports a delicately cut mother-of-pearl dial positioned on the three bridges. The blue alligator strap is complete with a buckle set with 18 diamonds.

GIRARD-PERREGAUX

GIRARD-PERREGAUX VINTAGE 1945 TOURBILLON – REF. 99880-52-000-BA6A

A horological treasure available in a limited edition of 50 pieces, the Girard-Perregaux Vintage 1945 Tourbillon with three gold bridges combines charm with mechanical mastery. The bridges and Art Deco-inspired case are fashioned in pink gold, which makes for a bold contrast with the matte anthracite-coated mainplate. The unique dark hue of the mainplate is the result of fusing precious materials and the use of innovative technology. It takes the masters at Girard-Perregaux one full week to achieve the perfect finish on this piece. The tourbillon mechanism requires special attention, as its 72 components must fit in a diameter of just 12mm.

GLASHÜTTE ORIGINAL

SENATOR TOURBILLON – REF. 94-03-04-04-04

The Senator Tourbillon is executed in white gold. The 42mm case of this timepiece features a slim bezel that serves as an elegant frame for the fine lacquered gray grained dial. The dial visuals feature Roman numerals and a classic railroad chapter ring engraved in the surface, which subsequently receives an elegant silver inlay. The distinctive Panorama date is presented in white on a dark ground beneath 12:00 and the poire hands in white gold enhance the transcendent elegance of the design. Taking pride of place on the dial is the superb flying tourbillon with small seconds at 6:00 in harmonious counterpoint to the Panorama Date.

GUY ELLIA

TOURBILLON MAGISTERE

Designer Guy Ellia's famously skeletonized mysterious-winding tourbillon is now available in rose gold. The Time Square Tourbillon Magistère's Caliber TGE 97 was created by Swiss manufacture Christophe Claret and features a 110-hour power reserve, one-minute tourbillon, a bottom plate and bridges. The movement beats at 21,600 vph within an entirely hand-chamfered cage.

GUY ELLIA

TOURBILLON MAGISTERE II

Guy Ellia has renewed its collaboration with Christophe Claret to present their latest timepiece enhanced with a tourbillon escapement: the Tourbillon Magistère II. Inside its distinctive case, an elongated rectangle (44.2x36.7mm), the skeleton movement features mysterious winding and features an "invisible" dial, whose main hands display the hours and minutes while seeming to float in space. In fact, the technical genius of this watch, produced in a limited edition of 12 pieces for each version, lies in the inversion of its bridges compared to a classic tourbillon timepiece. Hence the location of the winding crown at 6:00 and the use of the crown positioned at 12:00, while maintaining pressure on the security pushbutton at 10:00 to set the time. This piece is for lovers of fine watchmaking in search of the unusual.

GUY ELLIA

TOURBILLON MAGISTERE TITANIUM

Designer Guy Ellia presents the titanium version of his famous skeletonized mysterious-winding tourbillon. The Time Square Tourbillon Magistère's Caliber TGE 97 was created by Swiss manufacturer Christophe Claret and features a 110-hour power reserve, one-minute tourbillon, and a titanium bottom plate and bridges. The movement beats at 21,600 vph within an entirely hand-chamfered cage.

GUY ELLIA

TOURBILLON ZEPHYR

With a movement created by Swiss manufacture Christophe Claret, the Tourbillon Zephyr symbolizes transparency with excellence. Caliber GES 97 features a 110-hour power reserve and winding ring set with 36 baguette-cut diamonds (1.04 carats), or engine turning and one-minute tourbillon movement. The latter beats at 21,600 vph within an entirely hand-chamfered cage. The convex case with 950 platinum sides has a transparent back cover and hearth engraving. Mounted on an alligator strap with a folding buckle, this model is also available in platinum and titanium.

HUBLOT

BIG BANG TUTTI FRUTTI TOURBILLON PAVE – REF. 345.PO.2010.LR.0906

Elegantly feminine and full of joy, the Tutti Frutti Tourbillon Pavé injects luxurious delight into the queen of complications. Visible on the lower half of the dial, bordered by the romantic shimmer of 50 sparkling diamonds, the watch's flying tourbillon exposes the hand-wound mechanical heart of this exquisite timepiece. Framing the enchanting exhibition, a 41mm 18K red-gold case, set with 198 diamonds, sets the stage for its pavé-set bezel to amaze with the spectacle of 48 baguette orange sapphires.

HUBLOT

CLASSIC FUSION BLACK CERAMIC SKELETON TOURBILLON – REF. 505.CM.0140.LR

A striking fusion of past and present, of homage and modernity, the 45mm Classic Fusion Black Ceramic Skeleton Tourbillon perfectly combines the brand's passion for the avant-garde with its profound respect for a historic complication. Playing off the darkness of the case and black alligator strap, the watch's skeletonized dial reveals Hublot's visionary craftsmanship, exposing the meticulous artistry of the hand-wound MHUB6010.H1.1 caliber's splendid tourbillon at 6:00. Drawing on the rich history of the house, the polished hour and minute hands are designed to evoke the aesthetic sensibilities of the manufacture's earliest creations.

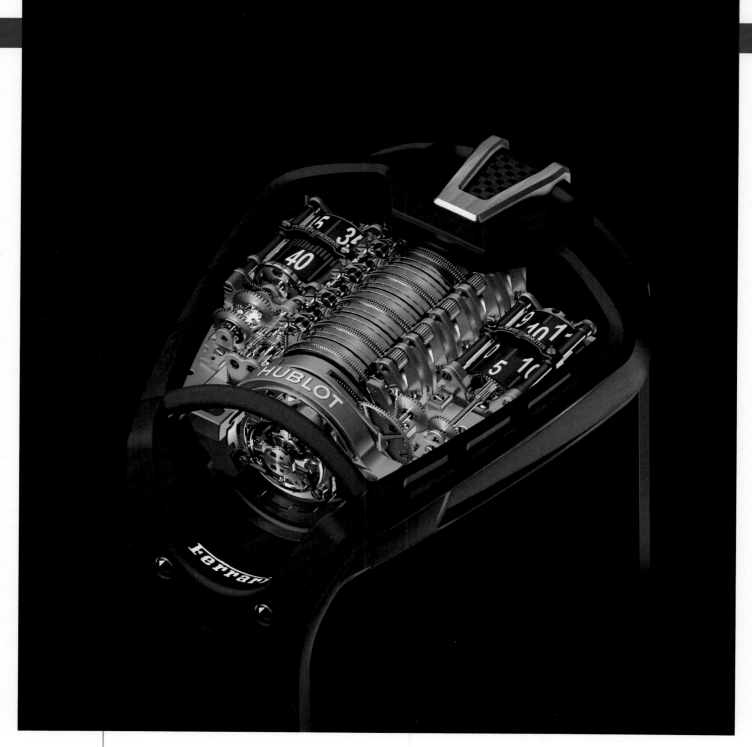

HUBLOT

MP-05 "LaFERRARI" – REF. 905.ND.0001.RX

The MP-05 "LaFerrari" fascinates with its spectacular design. Boasting 637 components and 11 series-coupled barrels arranged in the form of a spinal column, the 50-piece black PVD titanium limited edition timepiece boasts an unbelievable 50-day power reserve: a world record for a hand-wound tourbillon wristwatch. Complementing the display of hours and minutes to the right of the barrels, and power reserve to their left, via anodized black aluminum cylinders, the MP-05 "LaFerrari" features a generously sized 14.5mm suspended vertical tourbillon cage, upon which an additional aluminum cylinder is fastened to indicate the seconds.

IWC

INGENIEUR CONSTANT-FORCE TOURBILLON – REF. IW590001

Inside a 46mm case in platinum and ceramic, the 94800 hand-wound caliber, aided by its double-barrel construction, drives a 60-second tourbillon with integrated constant-force mechanism. Visible at 9:00 beneath a black bridge and slender seconds hand, the tourbillon preserves a constant amplitude of the balance thanks to an assembly that allows the escapement to be precisely uncoupled from the gear train. On the textured black dial, above the 96-hour power reserve indicator between 4:00 and 5:00, a highly realistic double-moon display with visible craters, achieved via a 3-D laser process, indicates the phases of the moon and countdown to the next full moon for both the Northern and Southern Hemispheres.

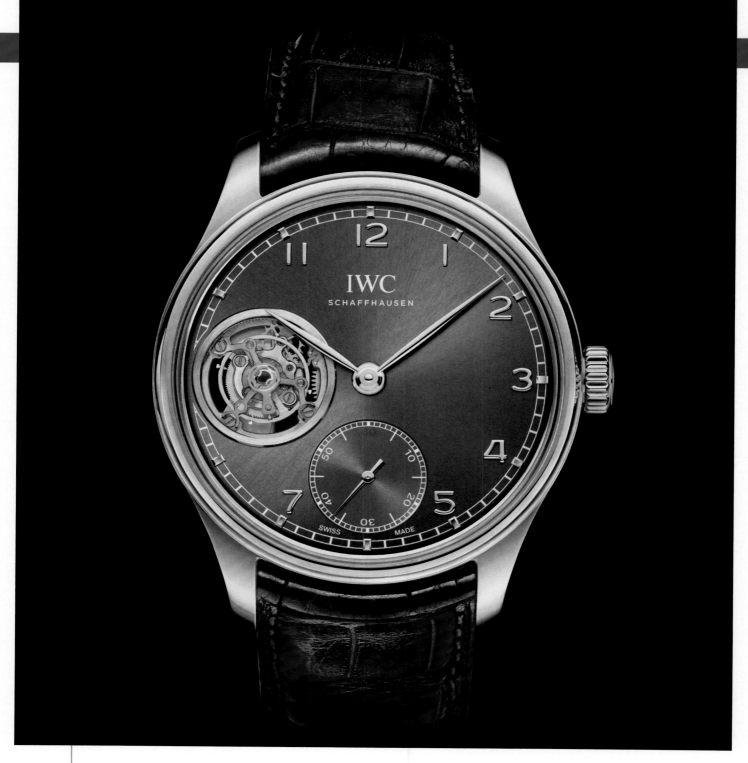

IWC

PORTUGUESE TOURBILLON HAND-WOUND – REF. IW546301

Revealed at 9:00 through an opening in the slate-colored dial, the 98900 caliber's 64-component flying tourbillon brings a taste of the 28,800 vph movement's micro-mechanical intricacies to the understated watch. In addition, the organ's rich golden tones perfectly contrast the sheen of the 43mm case's 18K white gold. The hand-wound timepiece, complete with small seconds at 6:00, reveals its movement's finely decorated nickel-silver three-quarter bridge through a sapphire caseback and is worn on a dark brown Santoni alligator leather strap.

JACOB & CO.

BRILLIANT FLYING TOURBILLON BLUE SAPPHIRE – REF. 210.543.30.BB.BB.1BB

Invisibly set with 160 baguette blue sapphires, the dazzling dial of this piece showcases the one-minute flying tourbillon carriage at 6:00 complete with the circular satin-brushed Jacob & Co. upper bridge. The Brilliant Flying Tourbillon Blue Sapphire is powered by the hand-wound JCBM01 caliber with a power reserve of 100 hours. Oscillating at 21,600 vph, the movement comprises 206 components and 29 jewels. Similar to the dial, the 47mm 18K white-gold case is invisibly set with 404 baguette blue sapphires as well. There is a total of 404 baguette blue sapphires at a weight of 36.75 carats for the lucky individual who wears this one-of-a-kind piece.

JACOB & CO.

EPIC X DIAMOND TOURBILLON – REF. 550.500.30.BD.BD.1BD

Worn on an openworked "honeycomb" rubber strap, this 18K white-gold timepiece is invigorated by 118 baguette-cut diamonds on its case and vertical bridges. The 44mm wristwatch is powered by the hand-wound JCAM03 caliber, distinguished by the vertical alignment of its balance and barrel with sliding clamp system. The hours and minutes are indicated via skeletonized hands at the center of the dial.

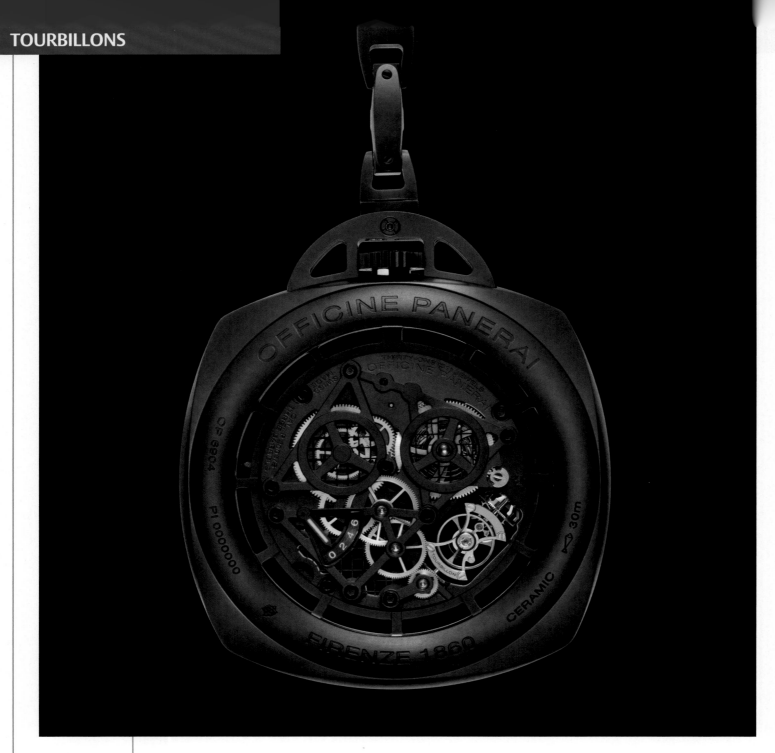

PANERAI

POCKET WATCH TOURBILLON GMT CERAMICA – 59MM – REF. PAM00446

The tourbillon mechanism, in its spectacular and sleek P.2005/S skeletonized version is the distinguishing feature in a timepiece of very high technical content: the Pocket Watch Tourbillon GMT Ceramica – 59mm, a model that pays tribute to Panerai's long history of watchmaking. The pocket-watch played an integral role in the brand's history, and this model echoes the past with innovative technical content of contemporary haute horology. In this Panerai original, the tourbillon cage contains the balance and escapement, rotating every 30 seconds (instead of the traditional 60) on an axis perpendicular to that of the balance.

PARMIGIANI

OVALE TOURBILLON – REF. PFH750-1003800

Powered by the hand-wound PF 501 skeleton caliber, this oval-shaped 18K rose-gold timepiece reveals, at 6:00 and 9:00, the detailed construction of its 237-component heartbeat. The tourbillon, admired on the lower half of the black mother-of-pearl dial, completes its full revolution every 30 seconds, while a golden retrograde semi-circle at 12:00 indicates the movement's remaining power reserve. Limited to 30 pieces and individually numbered on the caseback, this finely decorated timepiece boasts two series-coupled barrels and is worn on a black Hermès alligator strap.

PARMIGIANI

PERSHING SAMBA MADEIRA – REF. PFH551-3107000

This titanium wristwatch with 18K rose-gold bezel, individually numbered and engraved on its caseback with "MJF," "UNIQUE MODEL," and "BRAZIL," showcases its hand-wound movement's 30-second tourbillon on a dial adorned with "Gibson and Brazil" décor in wooden marquetry. Boasting a hand-driven power reserve indicator at 12:00 to complement the hours, minutes and central seconds, this 21,600 vph timepiece is expertly embellished to evoke the sights and sounds of its rhythmic Brazilian inspiration.

PARMIGIANI

PERSHING TOURBILLON ABYSS – REF. PFH551-3100600

Emerging through the surface of an undulating deep sea-blue dial, this 45mm timepiece's 30-second tourbillon provides a glance into the delicate micro-mechanical craftsmanship of the PF 510 hand-wound caliber with two series-coupled barrels, decorated with Côtes de Genève. Housed in a titanium case with 18K rose-gold bezel, this 30-piece limited edition is elegantly worn on an Hermès alligator strap with titanium folding buckle and indicates the hours, minutes, and central seconds, as well as the power reserve via a hand-guided retrograde display at 12:00.

PIAGET

PIAGET EMPERADOR COUSSIN XL TOURBILLON – REF. G0A37039

The world's thinnest automatic-winding tourbillon owes its title to the house's 1270P caliber. Measuring a mere 10.4mm in thickness, this new interpretation of Piaget's aesthetic codes possesses an architecture as spectacular as the exposed movement that powers the watch's heartbeat. Through a laser- engraved sapphire crystal, the wearer may admire an offset white-gold micro-rotor lined with 107 brilliant-cut diamonds (0.2 carat) as well as an imposing tourbillon cage, offering a veritable third dimension to a dial that features a delicate laser-guilloché craftsmanship and beautifully tapered hour markers. The case on this piece makes the model not only remarkable for watch aficionados, but for haute jewelry enthusiasts as well. The 18K white-gold case is set with over 1,000 diamonds weighing in at approximately 16.5 carats.

RICHARD MILLE

RM 021 TOURBILLON 'AERODYNE'

Continuing the expansion and application of truly unique materials to watchmaking, the RM 021 Tourbillon 'Aerodyne' is the first watch created with a composite baseplate utilizing a titanium exterior framework in combination with honeycombed orthorhombic titanium alumide and carbon nanofiber. This material was originally the subject of research by NASA for application as a core material of supersonic aircraft wings. This tourbillon displays hours, minutes, indicators for power reserve (circa 70 hours), torque indicator and function indicator for winding, neutral, and hand-setting operations. The PVD-treated variable inertia balance wheel with overcoil spring and ceramic endstone for the tourbillon cage offers optimal chronometric results and excellent long-term wear capability.

RICHARD MILLE

RM 38-01 BUBBA WATSON TOURBILLON G-SENSOR

For 2014, the famed Florida golfer returns with RM 38-01, a new golf watch with an integrated G-sensor mechanism that Richard Mille and his engineers were able to develop for the golf world. This mechanism is built within the heart of the hand-wound tourbillon movement, which has a power reserve of 48 hours. The G sensor at 12:00 records the force generated by one of Bubba's swings, specifically during the finish. RM 38-01 can sense an acceleration of up to 20 G's, necessary to record Bubba Watson, renowned for his swing, which is capable of projecting a ball more than 310 km/h. Resetting to 0 is easily accomplished by depressing the pusher at 9:00.

RICHARD MILLE

RM 36-01 TOURBILLON COMPETITION G-SENSOR SEBASTIEN LOEB

The new RM 36-01 G-sensor Tourbillon Sébastien Loeb is the result of collaboration between Richard Mille and the renowned F1 driver. This novel tourbillon movement has an indicator at 2:00 that displays the status of the circa 70-hour power reserve, a function selector at 4:00 and a G-force sensor reset in the middle of the round sapphire crystal front. This new version of the G-force sensor can now be rotated manually to align in different directions, enabling drivers to visualize the lateral deceleration as well as longitudinal forces found respectively in large curves and acceleration and braking in straightaways.

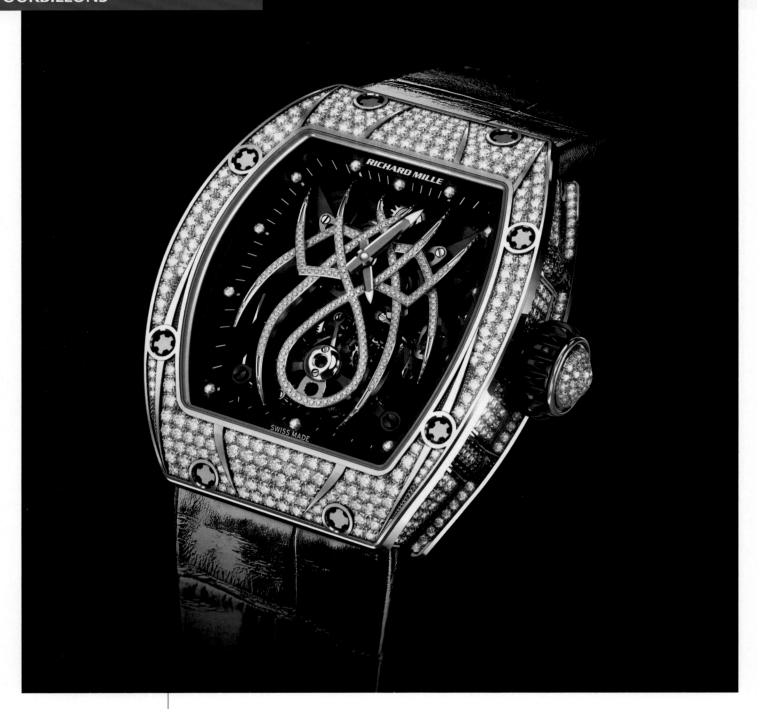

RICHARD MILLE

RM 19-01 TOURBILLON NATALIE PORTMAN

Natalie Portman, the famed American actress and partner of the brand since 2011, unveils the first watch designed and created in collaboration with Richard Mille. This manual-winding tourbillon is built on a black rhodium-plated, 18K white-gold baseplate, a first for Richard Mille. The spider within the movement, also in 18K white gold and delicately set with diamonds, forms an integral part of the caliber, with the abdomen of the spider actually supporting the bridges of the tourbillon, and its legs supporting the two winding barrels. Released in a limited edition of 20 pieces, this model is available in 18K white or red gold with a gem-set case.

RICHARD MILLE

RM 51-01 TOURBILLON TIGER AND DRAGON MICHELLE YEOH

This novel artistic creation celebrates Michelle Yeoh's 2014 film *Crouching Tiger, Hidden Dragon II*. Yeoh wanted to focus on the symbolic themes of the tiger and the dragon, with both creatures clutching the tourbillon movement. Created in 3N red gold, the figures are carved and incised entirely by hand using special miniature tools. With a power reserve of 48 hours, indicated by a red line between 10:00 and 11:00, the grade-5 titanium movement has a torque-limiting crown that protects the watch from possible overwinding. This piece is available in a limited edition of 20 pieces in 18K white or red gold.

RICHARD MILLE

RM 59-01 TOURBILLON YOHAN BLAKE

The 24-year-old Jamaican Yohan Blake, multiple medalist at the London Olympics in 2012, is a member of the very exclusive circle of the world's fastest sprinters and a partner of the brand. The RM 59-01 created for him has visually impressive, dynamically shaped bridges colored in green and yellow, machined from anticorodal aluminum Pb109, an alloy of aluminum, magnesium, silicon and lead. The skeletonized, manual-winding 21,600 vph tourbillon movement has a power reserve of approximately 50 hours. It offers an optimal performance/regularity/power reserve ratio, provided by the combination of grade-5 titanium with anticorodal aluminum Pb109, optimized gear trains and a variable-inertia balance. The asymmetrical case design is made from a translucent composite injected with millions of nanotubes in carbon, one of the strongest materials currently known to man.

ROGER DUBUIS

EXCALIBUR SKELETON DOUBLE FLYING TOURBILLON – REF. RDDBEX0395

The iconic case of the Excalibur 45 collection demonstrates the appeal of the Excalibur identity codes. At the same time, it provides a new presentation for the skeletonized double tourbillon within which is of outstanding quality and original design. By reducing the plates and bridges to their absolute minimum, Roger Dubuis has created a spacious yet highly technical mechanism, the RD01SQ movement. The Geneva watchmaker thus presents a design with an open heart and a top-flight movement, whose thinner case gives the signature codes a new dynamism.

STÜHRLING ORIGINAL

SATURNALIA TOURBILLON

Combining the refined aesthetics of a stamped guilloché-like pattern with the energetic revelation of a skeletonized architecture, this 45mm stainless steel timepiece frames its imposing open-heart tourbillon with an elegant and sophisticated stage. A 24-hour indicator at 3:00 joins an off-centered presentation of the hours and minutes for a clear and unambiguous reading of the time and day/night status. At 9:00, a snailed subdial informs the wearer of the ST-93961 caliber's remaining power reserve. The engraved scrollwork décor of the movement, exhibited through the tourbillon's large opening at 12:00, adds a final touch of watchmaking finesse to a timepiece that bestows upon its complicated movement a brilliant degree of aesthetic know-how.

STÜHRLING ORIGINAL

METEORITE TOURBILLON

Emerging from the stunning visual appeal of this timepiece's raw meteorite dial, the ST-93301 caliber's open-heart tourbillon grants its owner a glimpse into the fine inner workings of the movement's sophisticated construction. Two leaf-shaped hands indicate the hours and minutes against a sequence of individually applied Roman numerals that echo the rich gold tone of the watch's 44mm stainless steel case, while the stainless steel movement, decorated with Côtes de Genève, may be enjoyed through an exhibition caseback. Limited to 400 pieces, the Meteorite Tourbillon possesses a power reserve of 36 hours and oscillates at a frequency of 28,800 vph.

TAG HEUER

CARRERA MIKROTOURBILLONS – REF. CAR5A51.FC6331

The Carrera MikrotourbillonS is not only the world's fastest tourbillon: it is the first ever tourbillon on a 1/100th second chronograph that can be started and stopped. The MikrotourbillonS has two rotating tourbillon mechanisms visible on its dial face. The first beats at 4Hz (28,800 vph) and controls the watch. The second, the world's fastest tourbillon, controls the 1/100th second chronograph and runs at 50Hz, meaning it beats at 360,000 vph and rotates at a dizzying five seconds per revolution, or 12 times a minute. Another technical feat: the cage-free mechanism can be started and stopped thanks to its dual-chain architecture, which incorporates one kinematic chain for the watch and a second for the chronograph.

VACHERON CONSTANTIN

MALTE TOURBILLON – REF. 30130/000R-9754

Powered by the Vacheron Constantin 2795 caliber, the Malte Tourbillon is stamped with the Poinçon de Genève. This model, which was developed and manufactured entirely in-house, has a power reserve of approximately 45 hours and oscillates at 18,000 vph. The glistening 18K 5N pink-gold case is 38x48.24mm and acts as the perfect frame for the stunning silvered sand-blasted dial within. The tourbillon carriage with small seconds is at 6:00 and is accompanied by 18K 5N pink-gold hour markers, Roman numerals and a black painted minute-track.

VACHERON CONSTANTIN

PATRIMONY TRADITIONNELLE TOURBILLON 14 DAYS – REF. 89000/000R-9655

The new Vacheron Constantin Caliber 2260 beating at the heart of the Patrimony Traditionnelle 14-day Tourbillon is simply unrivaled. In order to provide a full 14-day power reserve, the new caliber is equipped with four barrels mounted in coupled pairs. The architecture of the Caliber 2260 comprises 231 components and two large bridges instead of the three appearing on older Vacheron Constantin tourbillons. In addition to the tourbillon combined with small seconds at 6:00, the movement also drives the slightly off-centered hour and minute functions, as well as the power reserve indicator. The new Patrimony Traditionnelle 14-day Tourbillon faithfully reflects the design codes that have forged the reputation of this line imbued with a sense of purism and rigorously high standards.

ZENITH

PILOT MONTRE D'AÉRONEF TYPE 20 TOURBILLON – REF. 87.2430.4035/21.C721

This titanium timepiece with 18K rose-gold bezel plays with asymmetry to achieve a sophisticated architectural choreography reminiscent of vintage aviation instruments. Powered by the self-winding 4035 mechanical movement, with a power reserve of 50 hours, an intricately constructed tourbillon with integrated small seconds pierces the matte-black dial in its upper left quadrant. The date is positioned around the carriage of the gravity-regulating gyroscopic spectacle, cleverly displayed through a revolving aperture. Oscillating at 36,000 vph, the 381-component caliber also drives a chronograph with 30-minute and 12-hour counters.

Chronographs, Split-Seconds and Flyback Chronographs

A. Lange & Söhne
Double Split

Jean-Marc Jacot

CEO of Parmigiani

"The brand is building its history here and now, in step with our time."

Michel Parmigiani, an expert in watch restoration, founded the company that bears his name in 1996, with the support of the Sandoz Family Foundation. **THIS RECENT HISTORY GRANTS THE BRAND CONSIDERABLE LIBERTY, BACKED BY PERFECT MASTERY OF SEVERAL CENTURIES OF WATCHMAKING SAVOIR-FAIRE.** Founding member of the highly demanding Fleurier Quality Foundation (FQF) label, Parmigiani equips all of its watches with movements created internally. In the years to come, CEO Jean-Marc Jacot predicts more and more original products, in line with the expectations of the brand's clients.

What is Parmigiani's most distinguishing characteristic?

Our Fleurier production facility is one of the few to design, produce and assemble all the components of a watch: from the dial to the hands, from the gears to the balance spring. Thanks to the companies from the watchmaking center of the Sandoz Family Foundation, Parmigiani Fleurier enjoys total independence.

In addition, we are a young brand, founded upon the genius of an expert in watchmaking. Modernity and traditional savoir-faire are our signatures. For example, our finishings pay homage to age-old techniques, but we also create designs that are like no other. Parmigiani timepieces evolve within a very characteristic style that nonetheless affords a wide range of possibilities—we don't do "mono-products" at Parmigiani. Our collections include simple, classic models such as the Tonda 1950, but also high-tech watches such as the Bugatti. The brand is building its history here and now, in step with our time.

◀ Tonda Metrographe

Jean-Marc Jacot,
CEO of Parmigiani

^ Ovale Pantographe

^ Pershing Tourbillon Abyss

v Bugatti Super Sport

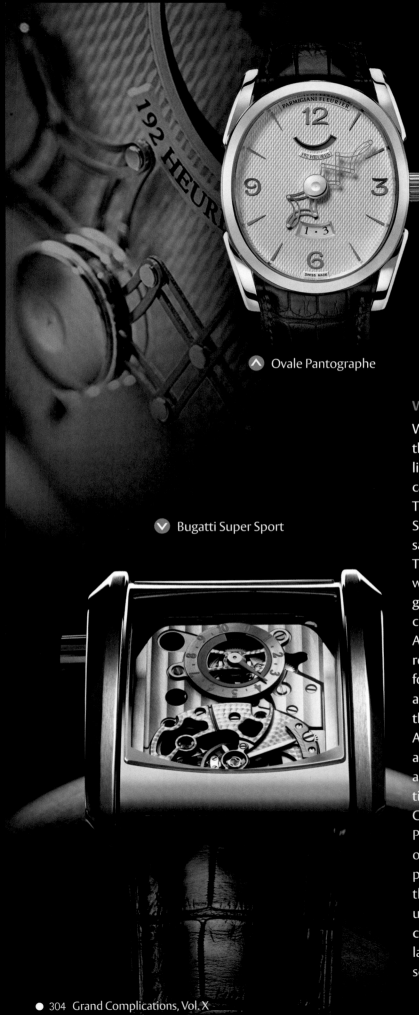

What does Parmigiani's product range comprise?

We actually have three "watchmaking fields" united under the name Parmigiani. The first, the Collection, comprises two lines: Tonda, with round cases, and Kalpa, with tonneau-shaped cases. Their original calibers are part of the brand's foundations. These models are offered for between 10,000 and 50,000 Swiss francs, and allow for multiple interpretations. For the same sector, we added the new Ovale line in 2013.

Then we have the creations from our haute horology workshops, where our master watchmakers construct our grand complications. Each of them is responsible for a watch, creating and adjusting it with an obsessive focus on perfection. Among these wonders, a minute repeater, for instance, requires 300 hours of work. A Westminster chime, with four gongs, requires 600. Of course, our watchmakers have also mastered one of the most spectacular complications, the tourbillon. In 2013 they created the Pershing Tourbillon Abyss, with 237 components and a tourbillon at 6 o'clock, all in a wide 45mm case. And finally, we create one-of-a-kind pieces that are absolutely exceptional. These are timekeeping objects, clocks or even automatons. For the Chinese Year of the Dragon, we created "The Dragon and the Pearl of Wisdom." The dragon makes a complete revolution over the course of one hour, in pursuit of an incandescent pearl. It escapes from the dragon's claws six times an hour, the gong's chime marking its flight. This work of art functions using a complex mechanism developed by Parmigiani. Its construction required 1,000 components and 5,800 hours of labor by numerous artisans, including a goldsmith, gemologist, sculptor and watchmaker.

What are you presenting this year?

In January, at SIHH, we presented Métro, a new collection aimed at a younger audience. A new interpretation of the Tonda, it is priced slightly lower, between 8,000 and 14,000 Swiss francs. Of course, it maintains the brand's signature codes and high-end quality. Concurrently, over the longer term, we are working on constructing an unusual caliber. It is a tremendous undertaking.

How has the brand evolved over the last five years?

We have worked relentlessly to reinforce the pillars of our brand: our industrial facilities, our products, our marketing and our distribution. The watchmaking center of the Vaucher Manufacture now guarantees us a complete production facility. On an aesthetic level, we've clarified our latest developments with precise design codes that recur throughout our watches, just as we have done in our marketing communication.

In terms of distribution, ten Parmigiani Fleurier retail stores have opened around the world. Truthfully, after the 2009 crisis, most independent distributors went out of business. It became practically impossible to find retailers that were capable of developing a brand like ours, especially over the long term. We didn't have a choice: we had to open up our own boutiques. This was—and still is—a pretty major investment, but it had to be done. This new expansion is already starting to get results!

What is your current distribution?

Once we have set up our distribution in Japan, we will control 90% of our global distribution. Parmigiani Fleurier is present in all the most important markets, about 50 countries in total. Currently, our watches are available in almost 250 points of sale. We'd like to raise that number to 350. Improving our brand's visibility also depends on this increased presence on the market.

In which country does the Parmigiani brand perform particularly well?

We work very successfully in Switzerland, and in France, where we repositioned the brand very well in 2013. We are very popular in Asia and the United Arab Emirates, as well as North America.

How would you sum up 2013?

We had a record year! We sold close to 6,000 watches. This is the result of the work we did to reposition the brand, working with the best retailers over 18 months. We have strengthened the existing bond of confidence with the distributors. Retailers did a better job of selling our watches, sometimes seeing considerable growth. Parmigiani Fleurier is now an established brand.

What growth do you foresee in 2014?

We are counting on a realistic goal of 15% growth per year in order to sustain it under the right conditions. Eventually, we wish to produce between 9,000 and 11,000 watches. To reach that point, we will improve and increase the number of points of sale, as well as the turnover of inventory. Barring another financial or economic catastrophe, we will reach this goal.

> The Dragon and the Pearl of Wisdom

Chronographs

The chronograph is one of the stars of 21st century watchmaking and one of the most on-trend complications. This unprecedented success is doubtlessly due to its practical utility, but also and above all to its technical and sporting touch that is so in sync with today's dynamic lifestyle. In response to this surge of enthusiasm, brands are offering an ever-increasing number of innovations in terms of design, materials, displays, mechanisms and performances. We are witnessing a spate of split-second chronographs (the most sophisticated form of this complication), along with other specialties such as mono-pusher, single-counter or regatta chronographs—and of course a number of exclusively feminine models. Another strong trend is that more and more brands are creating their own mechanical chronograph movements, instead of sourcing them as they used to do from the same set of "motor" suppliers.

THE PATERNITY OF THE CHRONOGRAPH WAS LONG ASCRIBED TO FRENCH WATCHMAKER NICOLAS RIEUSSEC, WHO IN 1821 CREATED AN INSTRUMENT THAT PLACED A DROP OF INK ON A DIAL AT THE BEGINNING AND END OF EACH MEASUREMENT.

However, the year 2013 brought a complete upheaval of this established order with the discovery of a 1816 timepiece signed by Louis Moinet—a compteur de tierces or a counter displaying sixtieths of a second by means of a central hand, seconds and minutes on two separate subdials, as well as hours on a 24-hour dial—complete with start, stop and reset functions controlled via two pushers.

The year 1838 saw the invention of the split-seconds chronograph (see "How does it work?"), and the first system to reset the hands to zero gained widespread popularity around mid-century. Now equipped with its three main functions (start, stop, reset), the chronograph would become very useful in several domains (science, technology, transportation, the military), and it is not by chance that its rise coincided with that of the industrial society and competitive athletics. During this period the chronograph would also gain several new scales for various measurements: tachometer (speed), telemeter (distance), pulsometer (pulse), etc.

The first wristwatch chronographs made their appearance in 1915. They were equipped, as were the pocket chronographs, with a single pushbutton for all three functions, usually housed in the crown. Breitling designed the now dominant chronograph configuration by adding the first independent pushbutton, and then in 1934, another one for resetting the chronograph. In 1937, the Dubois Dépraz manufacture presented a movement whose chronograph command operated using a cam system, less expensive and less difficult to produce than the traditional column wheel. The system would slowly be adopted by most brands.

In July of 1969, the Speedmaster, from Omega, would accompany American astronauts on their first visit to the moon. That year was also marked by the appearance of the first automatic chronograph movements. The movement developed by the Büren/Breitling/Heuer-Leonidas threesome in collaboration with Dubois-Dépraz was a cam movement featuring a modular construction, with an off-centered microrotor (oscillating weight); the one from Zenith, the famous El Primero, was an integrated movement with a column wheel in its central rotor. Thanks to its frequency of 36,000 vph, it was also the first chronograph movement capable of measuring times to one-tenth of a second.

The advent of quartz movements, which were much more precise, was nearly the deathblow for the mechanical chronograph, but the renaissance of mechanical horology in the late 1980s signaled its triumphant return. Let us remember that it is very rare for watch brands to design and produce their own chronograph movements. A large percentage of the chronographs on the market today house an ETA 7750 Valjoux movement, reworked to a greater or lesser degree and sold with differences in price that can be hard to justify... But the situation is beginning to change. The public, better informed on the subject than previously, is more than willing to pay a premium for a high-end chronograph, but on the condition that it is an exclusive movement—or at least somewhat original. The Swatch Group's announcement that it would stop selling ETA movements has also inspired competing brands to look for alternative solutions. For several years, we have been seeing an increase in the number of "in-house" chronographs featuring interesting innovations.

The chronograph functions using a complex system of levers, springs and wheels. The chronograph mechanism is "grafted" onto the movement's base. When the "start" pushbutton is pressed, the chronograph wheel joins the wheel for the seconds, and the hand starts ticking. When the chronograph is stopped, the chronograph wheel separates from that of the seconds, and the chronograph hand stops again. The reset function is effectuated by a system of a hammer on a spring and a "heart."

SPLIT-SECONDS CHRONOGRAPHS

Split-seconds chronographs are equipped with an additional seconds hand, which can be stopped independently to measure a "split" time or a reference time. They are recognizable by the additional push-button they bear. Activating the pushbutton is all that is needed for the split-seconds hand to catch up with the main chronograph hand, so that the two may continue running together.

COLUMN-WHEEL AND CAM MOVEMENTS

There are two kinds of chronographs with respect to the "control group" governing the start, stop and reset functions. The chronographs considered to be the most prestigious are equipped with a column-wheel, a complex organ that uses a system of slots and tabs to govern the movements of the levers. Most current mechanical chronographs are equipped with a cam system of simpler construction.

MODULAR AND INTEGRATED CONSTRUCTIONS

In modular constructions, the chronograph mechanism is totally independent of the base movement. It is mounted on an additional plaque affixed to the dial side of the movement. This solution allows the watchmaker to link a chronograph mechanism to an existing movement, without having to develop a new caliber. In integrated constructions, the chronograph components are placed directly on the movement, on the bridge side. Column-wheel chronograph movements are the result of integrated constructions.

FREQUENCY AND PRECISION

The precision of time measurement and display of a mechanical chronograph depend on the frequency of its regulating organ, i.e. the number of vibrations per hour (one oscillation equals two vibrations) of the balance-spring. The higher the frequency, the smaller units of time that can be measured.

Vibrations per hour	Frequency (in Hz)	Smallest measurable unit of time
18,000	2.5 Hz	1/5 of a second
21,600	3 Hz	1/6 of a second
28,800	4 Hz	1/8 of a second
36,000	5 Hz	1/10 of a second
360,000	50 Hz	1/100 of a second

FLYBACK FUNCTION

The flyback function allows the wearer to instantly start a new round of timing while the chronograph is already operating. For this, one need only activate the reset pushbutton, without stopping the hands first; all three chronograph hands will immediately return to zero to start another measurement.

JUMPING SECONDS

The flyback function allows the wearer to instantly start a new round of timing while the chronograph is already operating. For this, one need only activate the reset pushbutton, without stopping the hands first; all three chronograph hands will immediately return to zero to start another measurement.

CHRONOGRAPH AND CHRONOMETER

The word "chronometer" is often used incorrectly to designate a chronograph, but they are two very different things. A chronometer is a watch whose movement has received an official certificate attesting to its extreme precision. This certificate is awarded by the COSC (Swiss Official Chronometer Testing Institute) after two weeks of very rigorous testing. A chronograph may be a certified chronometer, but it is by no means certain that any given chronometer will also be a chronograph.

A Reinvented Classicism

Chronographs have long been distinguished by their rather classic aesthetic, with three small symmetrically placed subdials. This traditional look has now been swept aside by many models that reinvent the ways in which we measure time. The chronograph universe has also been marked by the appearance of new materials that allow the watches to become lighter, more robust or more technically sophisticated, as well as shaking up aesthetic codes.

Richard Mille
RM 11-01 Robert Mancini

Dedicated to former soccer player Robert Mancini, the RM 11-01 by Richard Mille features a chronograph system specifically designed for this sport. The dial is divided into match half-times, complete with an extra-time indication, and the flyback function serves to immediately restart timing of each half-time.

Breguet, *Type XXII 3880*

Equipped with a flyback function, the Type XXII 3880 by Breguet boasts an exceptional frequency of 10Hz (72,000 vph)—a feat made possible by the use of an escapement and flat balance-spring made of silicon.

Piaget
Polo FortyFive Chronograph

The Polo FortyFive Chronograph by Piaget houses an ultra-thin automatic manufacture movement—the House specialty—inside a satin-brushed case with polished gadroons. It also provides a 24-hour dual-time display at 9 o'clock.

Girard-Perregaux
Vintage 1945 XXL Chronograph

The Vintage 1945 XXL Chronograph by Girard-Perregaux houses an automatic movement entirely developed and produced within the brand workshops inside a cambered rectangular steel case.

Zenith, *El Primero Lightweight*

A high-tech version of its El Primero Striking 10th chronograph (with a central chronograph seconds hand that displays tenths of a second), the Zenith El Primero Lightweight combines a titanium and silicon movement with a carbon fiber case.

Tudor, *Heritage Chrono Blue*

A reinterpretation of a 1973 model, the Heritage Chrono Blue by Tudor is distinguished by its two counters appearing within trapeze-shaped frames, its blue aluminum bezel and its fabric strap picking up the dial colors.

Rolex, *Oyster Perpetual Cosmograph Daytona*

In 2013, Rolex celebrated the 50th anniversary of the Oyster Perpetual Cosmograph Daytona—one of the most legendary chronographs of all time—by launching a new version with a platinum case, chestnut brown Cerachrom bezel and glacier blue dial.

Jaeger-LeCoultre
AMVOX2 Chronograph Concept

On its AMVOX2 Chronograph Concept watch, Jaeger-LeCoultre introduced a small revolution in terms of controls. The traditional pushbuttons were replaced by a system in which the case middle pivots around a horizontal axis from 9 to 3 o'clock. To start or stop the chronograph, one need only press gently on the upper half of the crystal; resetting is done by a push on the lower half.

Breitling, *Navitimer 01*

Equipped with a slide rule serving to perform all operations relating to airborne navigation, the Navitimer chronograph launched in 1952 illustrates the close ties between Breitling and aviation. It is now equipped with a Manufacture Breitling movement.

Dubey & Schaldenbrand
Grand Dôme DT

On the tonneau-shaped Grand Dôme DT from Dubey & Schaldenbrand, the chronograph displays are arranged along a vertical axis, with minutes at 12 o'clock and hours at 6 o'clock.

Omega
Speedmaster Professional

The only watch to have been worn on the moon—hence its Moonwatch nickname—the famous Omega Speedmaster is presented in a version that stays true to the original, recognizable for its ultra-readable black dial and its tachometer.

Disc Displays

Varying the looks and methods of reading the time, watchmakers sometimes have a little fun by replacing the chronograph minutes hand (or even the hour or seconds counter hand) with an aperture and disc display—an excellent way to combine originality and readability.

> **Montblanc**
> *Nicolas Rieussec Open Date*
> *Silicium Escapement*

A tribute one of the pioneers of the chronograph, the Nicolas Rieussec Open Date Silicium Escapement from Montblanc displays the chronograph minutes and seconds using two rotating discs equipped with fixed pointers.

> **Cartier**, *Rotonde Central Chronograph*

On the Rotonde Central Chronograph from Cartier, the chronograph indications have the place of honor at the center of the dial, with a seconds hand and a counter that sweeps across an arc as a 30-minute totalizer.

> **Jaeger-LeCoultre**
> *Master Compressor Extreme LAB 2*

Designed to withstand even the most challenging conditions, the Jaeger-LeCoultre Master Compressor Extreme LAB 2 is distinguished by its patented digital counter and its very readable way of indicating the chronograph's jumping minutes.

> **Pierre DeRoche**
> *GrandCliff TNT Penta*

Equipped with an exclusive Dubois Dépraz movement, the GrandCliff TNT Penta uses five discs (hence its name) to display the large date and the small seconds as well as the 60-minute and 12-hour counters of the flyback chronograph.

Quick-Change Artist Chronographs

Trying to decide whether to purchase a straightforward watch or a chronograph? Prospective buyers no longer have to choose between these options, since several brands have invented ingenious mechanisms that modify the dial's appearance, allowing the wearer to switch from one to the other at will. This bit of magic enables them to offer two watches in one.

< Revelation
R03 Chronographe RS

The R03 Chronograph RS from the youthful brand Revelation is distinguished by its patented dial enabling it to switch from opaque black (with clearly readable chronograph counters) to a transparent version providing a full view of the movement—all with the turn of the bezel.

> Montblanc, *Metamorphosis*

On the spectacular Metamorphosis from Montblanc, a bolt at 9 o'clock causes the hours/minutes/seconds/date dial to disappear as if by magic, making way for the chronograph's minute counter, as the central seconds hand becomes that of the chronograph function.

< Valbray, *Oculus V.01 Chrono*

The Oculus V.01 Chrono model by Valbray features a dial base reminiscent of camera diaphragms. When the bezel is turned, the simple hours/minutes display gives way to an authentic chronograph dial.

Time for Women

A sign of the times, and a notable trend at the beginning of the third millennium, a growing number of chronographs are designed exclusively for feminine wrists. These sporty, elegant watches are often adorned in soft colors and graced with finely finished dials featuring materials such as mother-of-pearl and diamonds, following the lead of more formal watches. These beauties meld technical excellence and aesthetic refinement.

Patek Philippe, *Ladies First Chronograph Ref. 7071*

Launched in 2009, the Ladies First Chronograph Ref. 7071 from Patek Philippe was the first watch to house the brand-new in-house chronograph caliber CH 29-535 PS, which is classically constructed; it has since appeared in a masculine model.

Girard-Perregaux, *Small Column-Wheel Chronograph*

Girard-Perregaux offers a beautiful combination of technique and aesthetics on its automatic-winding Small Column-Wheel Chronograph for women, whose openworked dial features four large Arabic numerals of highly original design.

Breguet
Marine Chronograph for Lady

The Marine Chronograph for Lady by Breguet offers an original combination of a steel case, mother-of-pearl dial and rubber strap. It houses an automatic chronograph movement with a hand-decorated oscillating weight.

Chopard, *Imperiale Chronograph*

Immediately identifiable by its bracelet attachments reminiscent of ancient columns, the Imperiale Chronograph by Chopard comes in white gold, pink gold, steel, steel or blackened DLC steel versions, enhanced on some models by the sparkle of diamonds.

Blancpain
Large Date Chronograph

The Large Date Chronograph from the Blancpain Women collection associates a diamond-set bezel with a mother-of-pearl dial with an off-centered hour/minute indication. The sapphire caseback allows for admiration of an automatic movement equipped with a petal-shaped oscillating weight.

The Fashion for Single Counters

Eschewing the system of two small separate counters for the minutes and the hours, a growing number of watchmakers are opting for a single counter that displays both indications. The goal is to enable simpler, quicker reading of longer time measurements, just as one might read the time on the central dial.

> Pierre *DeRoche, SplitRock*

On its SplitRock, Pierre DeRoche emphasizes simplicity and ease of reading by proposing a chronograph with three concentric hour/minute/second hands, equipped with an exclusive Dubois Dépraz caliber.

^ IWC
Portuguese Yacht Club Chronograph

A new interpretation of an iconic IWC model, the Portuguese Yacht Club Chronograph displays the chronographs hours and minutes in a subdial at 12 o'clock. The chronograph also boasts a flyback function.

^ Panerai
Luminor 1950 3 Days Chrono Flyback Automatic

On the Luminor 1950 3 Days Chrono Flyback Automatic by Panerai, the customary chronograph minute counter is replaced by a large steel-toned instant-jump central hand.

> Omega, *De Ville Co-Axial Chronograph*

Offered in steel or red gold, Omega's De Ville Co-Axial Chronograph is distinguished by its single counter for hours and minutes at 3:00, as well as by its movement with exclusive co-axial escapement.

^ Jaeger-LeCoultre
Duomètre à chronographe

On the Duomètre à chronographe by Jaeger-LeCoultre, equipped with two independent mechanisms synchronized by the same regulating organ, the hour and minute counters are grouped together at 2:30, while a flying seconds hand at 6 o'clock allows measurements of one sixth of a second.

Single-Pusher Chronographs

In terms of specialty constructions, we are seeing an increase in single-pusher, or monopusher, chronographs. These pieces feature a sophisticated arrangement: the sole pushbutton handles the start, stop and reset functions, in succession.

⌃ Longines
Avigation Oversize Crown

Inspired by a 1920s Longines model, the Avigation Oversize Crown has all the signature features of pilot's watches, including a large crown that can be easily handled with gloves and that also serves as a chronograph monopusher.

⌃ Jaquet Droz
Complication La Chaux-de-Fonds Chrono Monopusher

The Complication La Chaux-de-Fonds Chrono Monopusher from Jaquet Droz is notable for being the first timepiece of its kind equipped with an hour counter as well as an off-centered hour and minute indication.

⌃ Jaeger-LeCoultre
Master Grande Tradition Gyrotourbillon 3 Jubilee

Jaeger-LeCoultre has equipped its Master Grande Tradition Gyrotourbillon 3 Jubilee with a mono-pusher chronograph featuring a subdial at 9 o'clock. The seconds are indicated by the central hand, and the minutes in an instant digital-display dual aperture.

⌃ Louis Vuitton
Tambour Twin Chrono

As a monopusher dual chrono-graph with a differential display, the Tambour Twin Chrono by Louis Vuitton enables simultaneous measurement of two separate times at 4 o'clock and 7 o'clock), while displaying the difference between them (at 12 o'clock). A useful function for competitions such as Match Racing regattas.

Christophe Claret, *Kantharos*

The monopusher Kantharos chronograph by Christophe Claret is equipped with a striking mechanism signaling any changes in the chronograph function on a cathedral-type gong. It also features a constant-force system visible on the dial.

Hundredth-of-a-second—or thousandth-of-a-second—timing

For many years, the display of time to one-tenth of a second reigned as the supreme accomplishment for a mechanical chronograph. A few watchmakers have taken up the challenge, daring themselves to break the sound barrier and present chronographs able to measure and display one one-hundredth and even one one-thousandth, of a second—notably using bi-frequency systems, with a regulating organ for the watch part and a separate regulating organ for the chronograph part. It is an admirable technical advance, even though the inevitable reaction time and delay in activating the pushbutton limits the usefulness of such a function in the field.

F.P. Journe, *Centigraphe Sport*

On the Centigraphe Sport from F.P. Journe, equipped with a single balance wheel beating at 21,600 vph, hundredths of a second are read off using a "flying" hand that makes a full rotation in one second. The tachometric scale can measure speeds of up to 360,000 km/h! Entirely crafted in aluminum alloy, this chronograph weighs just 55g.

TAG Heuer, *Carrera Mikropendulum*

To enable it to display hundredths of a second using a central hand, the Carrera Mikropendulum by TAG Heuer is equipped with a second regulating organ for the chronograph complete with magnetic system (no balance-spring) oscillating at 360,000 vph (50Hz).

Montblanc
TimeWriter II Chronograph Bi-Fréquence 1000

On its TimeWriter II Chronograph Bi-Fréquence 1000, Montblanc measures time to one-thousandth of a second with a second balance that "only" beats at 360,000 vph (50Hz). The "sweep seconds hand" makes a complete revolution around the dial in one second, and the thousandths of a second are displayed in an aperture at 12:00.

TAG Heuer, *Mikrogirder*

Equipped with an unusual regulating organ oscillating at 1,000Hz (7,200,000 vph), TAG Heuer's concept watch Mikrogirder measures time to 5/10,000 or 1/2,000 of a second. Its bi-frequency system uses two independent "chains," one for the watch and one for the chronograph.

Split-Seconds Chronographs

Split-seconds chronographs possess an additional seconds hand that can be stopped to measure an intermediate or spit time. This ultra-sophisticated clutch system is wide recognized as one of the most difficult complications to master—the equal of a tourbillon or a minute repeater—and the most prestigious brands are eager to offer it among their collections.

> **Blancpain**, *L-evolution Chronographe Flyback à Rattrapante*

Joining carbon fibers and cutting-edge technology, the L-evolution Chronographe Flyback à Rattrapante reflects Blancpain's engagement in the world of automotive competition. It is also equipped with a two-aperture large date display.

∧ **Richard Mille**
RM 056 Tourbillon Split Seconds Competition Chronograph Felipe Massa Sapphire

Dedicated to the Brazilian Formula 1 driver of the same name, Richard Mille's RM 056 Tourbillon Split Seconds Competition Chronograph Felipe Massa Sapphire is notably distinguished by its completely transparent case, carved from a solid block of sapphire.

∨ **Patek Philippe**
Ladies First Split Second Chronograph Ref. 7059

Patek Philippe paid tribute to women with the launch of its Ladies First Split Second Chronograph Ref. 7059, a single-pusher split-seconds chronograph powered by an ultra-thin movement that also drives several men's models.

> **A. Lange & Söhne**, *Lange Double Split*

The German brand A. Lange & Söhne presents a new innovation with its Lange Double Split, the first wristwatch chronograph with both a split-seconds mechanism and a split-minutes mechanism.

Regatta Chronographs

Models described as "regatta chronographs" possess a notable countdown system that marks off the last five or ten minutes before the race begins, before starting the timing operation as such. With the growing popularity of sailing competitions such as the America's Cup, these models are getting new wind in their sails.

< Corum, *Admiral's Cup AC-One 45 Regatta*
Recognizable by its twelve-sided bezel, the Admiral's Cup AC-One 45 Regatta from Corum has a countdown function that can be adjusted up to ten minutes via the crown, with a digital aperture-type display ensuring optimal readability.

∨ Panerai
Luminor 1950 Regatta 3 Days Chrono Flyback Automatic Titanio

In addition to its pusher-adjustable countdown system, the Luminor 1950 Regatta 3 Days Chrono Flyback Automatic Titanio by Panerai displays a tachometric scale in nautical knots serving to measure boats' speed.

> **Rolex**, *Oyster Perpetual Yacht-Master II*

Rolex has reaffirmed its place in the sailing world by launching the Oyster Perpetual Yacht-Master II, which boasts an ingenious bezel/crown/pushbutton system that follows the countdown using a large triangular hand.

∧ **Louis Vuitton**
Tambour Spin Time Régate

On the Tambour Spin Time Régate by Louis Vuitton, the five minutes of the countdown used in the America's Cup are indicated by five rotating cubes switching from red to black.

> **Omega**
Seamaster Diver ETNZ Limited Edition

A tribute to the challenger in the America's Cup 2013 (Emirates Team New Zealand), the Seamaster Diver ETNZ limited edition by Omega features a regatta countdown display at 3 o'clock. Its movement is fitted with a co-axial escapement.

A. LANGE & SÖHNE

1815 CHRONOGRAPH – REF. 402.032

The 1815 Chronograph houses an A. Lange & Söhne manufactured manual-winding L951.5 movement complete with a hand-engraved balance cock. One of the highlights among the numerous technical features is the flyback function with a precisely jumping minute counter, a device found in very few chronographs. The two symmetrically positioned subsidiary dials for the seconds and 30-minute counter emphasize the balanced geometry and the classic style of the watch. The pink- or white-gold case measures 39.5mm and is fitted with antireflective sapphire crystal on the front and back. The dial is solid silver with blued steel hands. The hand-stitched crocodile strap is secured with a solid gold buckle.

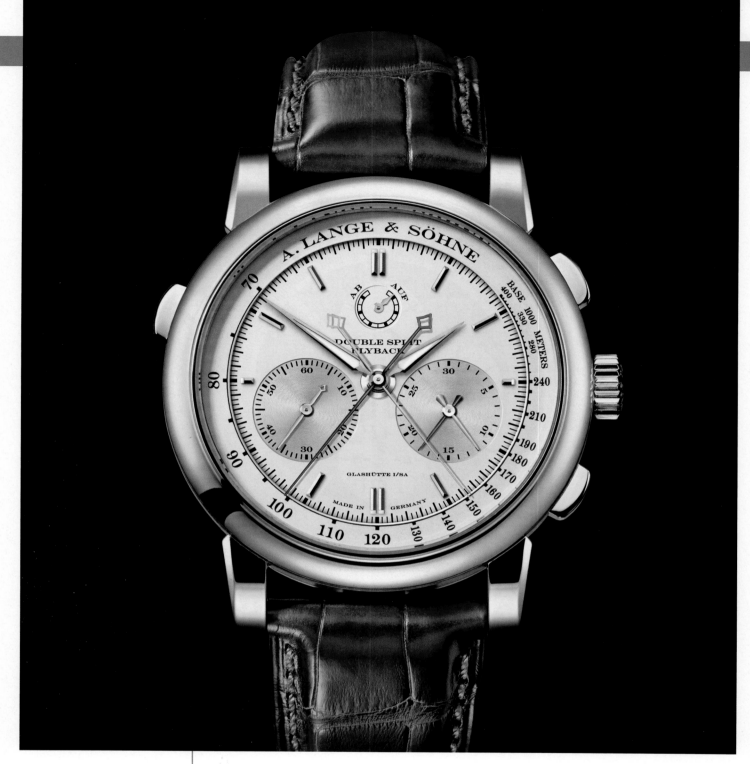

A. LANGE & SÖHNE

DOUBLE SPLIT – REF. 404.032

This model is powered by the manual-winding Lange L001.1 movement. The Double Split is a flyback chronograph with double rattrapante, controlled by classic column wheels. Included is a jumping chrono minute counter and rattrapante minute counter, flyback function with indication for hours, minutes, small seconds with stop seconds and power reserve indicator. The movement is encased in a 43mm rose-gold case with sapphire crystal on the front and caseback. The dial is two-tiered solid silver. The Double Split is also available in platinum.

AUDEMARS PIGUET

ROYAL OAK OFFSHORE CHRONOGRAPH – REF. 26401PO.OO.A018CR.01

The Royal Oak Offshore Chronograph 44mm is powered by Audemars Piguet's in-house, automatic-winding Manufacture caliber 3120, which is comprised of 280 parts and beats at 21,600 vph. The movement is enclosed in a 950 platinum case and includes a black ceramic bezel and crown, as well as pushbuttons and pushbutton guards. The chronograph features a date with the small seconds at 12:00. The pin buckle is also made of 950 platinum and the sapphire crystal caseback reveals the subtlety of its mechanism. This piece has anti-magnetic protection and is water resistant to 100m.

BLANCPAIN

CHRONOGRAPH GRANDE DATE – REF. 3626-2954-58A

The shimmering mother-of-pearl dial surrounded by a minute track reveals a two-part composition delineated by two waves of 17 variously sized diamonds. They ripple out from the center of the dial, along with the chronograph seconds hand. At 12:00, the off-centered time display featuring Roman numeral hour markers creates a dynamic counterpoint to the Arabic numerals on the chronograph counters. The double-disc large date display appears through twin apertures at 6:00.

BREGUET

CLASSIQUE CHRONOGRAPH – REF. 5287BR.12.9ZV

This timepiece has a 42.5mm case in white or rose gold, with a dial that displays a central chapter of hours and minutes and the seconds with a double hand at 9:00. The chronograph indications include a minute counter at 3:00 and the red central chronograph seconds hand. A tachometer scale to calculate speed discreetly surrounds the Roman numeral hour markers. The dial is engine-turned in a hobnail pattern in the center and in a circular barleycorn motif for the double seconds. Cross-hatching and snailing for the minutes counter complete the dial decorations.

BREGUET

MARINE LADIES CHRONOGRAPH – REF. 8827ST.5W.SMO

Introduced in a 34.6mm steel case, the Marine Ladies Chronograph is water resistant to 5atm. This model displays an ultra elegant dial in white mother-of-pearl, adding beauty to a functional chronograph. Pressing the chronograph button sets in motion the central seconds hand, 30-minute counter at 3 and 12-hour counter at 9. In addition to these functions, the central hours and minutes, small seconds and date are all powered by the caliber 550, an automatic-winding movement with a power reserve of 45 hours. This exceptional timepiece is available on either a metal bracelet or a turquoise leather strap.

BREGUET

TYPE XXII – REF. 3880BR.Z2.9XV

The emblematic Breguet Type XXII takes on an 18K rose-gold case. This fine material highlights the beauty of this unique timepiece, which houses a Breguet chronograph movement with silicon escapement and flat balance spring whose frequency has been raised to 72,000 vph, endowing the movement with exceptional regulating power. Its chronograph seconds hand effects a complete rotation of 30 seconds, making the timepiece's start function and readout twice as precise. At the heart of this technical exploit is the use of silicon, resulting in more lightweight mobile components and the prevention of lubrication problems generated by high frequencies.

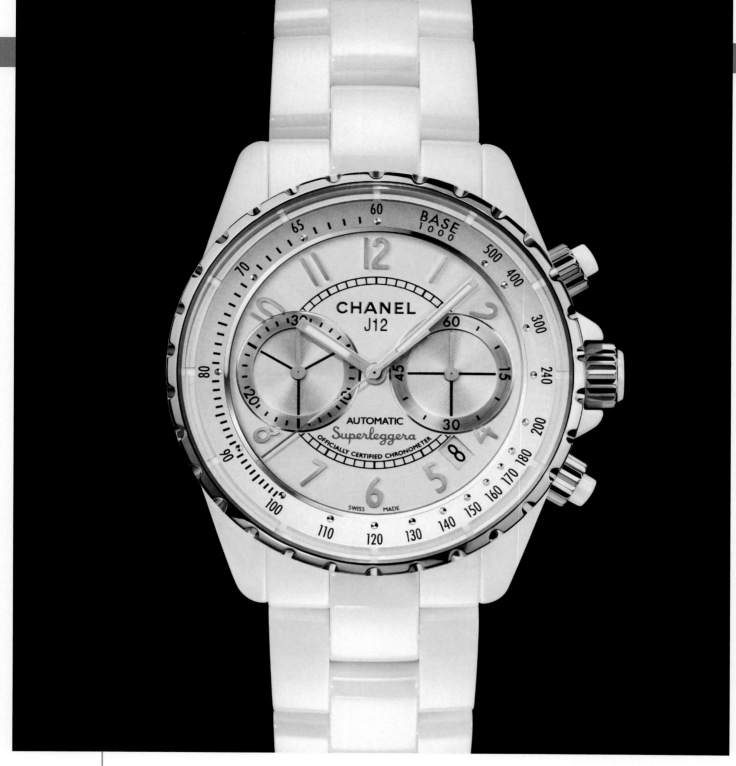

CHANEL

J12 SUPERLEGGERA – REF. H3410

White high-tech ceramic, highly scratch-resistant material. Self-winding mechanical chronograph; movement certified COSC (Swiss Official Chronometer Control); 42-hour power reserve. Functions: hours, minutes, seconds (counter at 3:00); date; chronograph (30 minute counter, central seconds) and tachymeter. Screw-down push pieces and crown. Steel case back engraved with red rubber "Superleggera". Steel triple folding buckle. Water resistant 200 metres. Diameter: 41mm. Also available in matte black high-tech ceramic.

CHANEL

J12 SUPERLEGGERA – REF. H3409

Since 2000, the J12 watch has embodied elegant and resolute masculinity. New models have been added to the shiny black collection inspired by these prestige automobiles, where the rich dark color has been treated to achieve an ultra matte black ceramic hue that is contemporary and intensely masculine. The sleek shape of the new J12 Superleggera watch, in matte black ceramic, reminds its wearer of speed, precision and performance. The elegance of its design, symbolic of automobile excellence, is matched perfectly by its high precision chronograph function.

Functions: hours, minutes, seconds (counter at 3:00); date; chronograph (30-minute counter, central seconds) and tachometer.

CHRISTOPHE CLARET

KANTHAROS – REF. MTR.MBA13.905

This monopusher chronograph with striking mechanism and constant force escapement asserts its own identity and reinvents the measurement of timing events. The automatic-winding chronograph's cathedral gong, visible at 10:00, audibly chimes with each change of function. In addition to this innovative complication, Kantharos is also equipped with a constant-force escapement at 6:00. This mechanism is a key component that considerably reduces timing variations. The energy delivered to the escapement remains constant from beginning to end of the power reserve. Precision is the overriding goal. The perpetually moving mechanism exercises a hypnotic visual effect, which may be admired beneath a meticulously chamfered sapphire bridge revealing the full extent of the master-watchmaker's know-how.

CHRISTOPHE CLARET

DUALTOW – NIGHTEAGLE – REF. MTR.CC20A.045

The Dualtow – NightEagle features a planetary-gear, monopusher chronograph with an original striking mechanism signaling start, stop and reset function. It also contains a tourbillon escapement and two spectacular rolling belt indicators running parallel on each side of the dial to indicate hours and minutes, the digital numerals evoking a cockpit instrument panel. This model draws inspiration from the ultra-high-tech world of stealth aircraft (particularly the American F-117 Night Hawk), with its smartly-angled architecture of tinted sapphire bridges covering the dial, combined with quasi-monochromatic black and gray features.

F.P. JOURNE

CENTIGRAPHE SPORT

The Centigraphe Sport belongs to the lineSport collection, a collection of watches of an astounding lightness, conceived specifically for advanced athletic activity. Inspired by an aficionado of F.P. Journe timepieces, the lineSport collection was designed to fulfill his wish for an ultra-light sports watch housing an authentic haute horology movement. Measuring time to 1/100 of a second, the Centigraphe Sport is the perfect match for athletic activities in which every second counts. F.P. Journe undertook intensive research to find an ultra-light and -resistant material to craft the house's exceptional calibers, while maintaining strict standards of haute horology. The Centigraphe Sport thus presents the very first F.P. Journe wristwatch entirely—movement, case and bracelet—made of aluminum alloy in the search for absolute comfort and extreme lightness. The watch in its totality only weighs 55g.

FRANCK MULLER

CHRONOGRAPHE AUTOMATIQUE MARINER

The Chronographe Automatique Mariner's radiant rose-gold case comes complete with a black crown and pushbuttons. These dark external parts not only match the natural rubber strap but the black sunray guilloché dial as well. The dial is spruced up by royal blue appliqués and white Arabic numerals and hands. The color scheme and compass décor of the dial give the model a vintage navigational feel. The chronograph is to 1/5 of a second with a 30-minute counter at 3:00 and a central seconds hand. The small seconds at 9:00 and date at 6:00 also guarantee exactitude for the Mariner's owner.

FRANCK MULLER

CONQUISTADOR GRANDPRIX CHRONOGRAPH

Bright, bold and dashing, the Conquistador Grandprix Chronograph shows Franck Muller's adventurous side. With an automatic-winding FM 700 caliber, a power reserve of 48 hours and a platinum 950 oscillating mass, this model is as powerful on the inside as it is on the outside. The black dial has a flame pattern with Arabic numerals painted in red relief. The titanium and ergal curved case of this Conquistador model is complemented by a black alligator strap with red stitching. Black and red make for an exceptionally strong-looking timepiece that is extremely functional as well. This can be attested for by its ultra-reliable bi-wheel chronograph.

GUY ELLIA

JUMBO CHRONO

The Jumbo Chrono, the first round men's watch created by Guy Ellia, is based on an interesting technological approach demonstrating that it is possible to combine aesthetic balance with masculine detail and a fine mechanical movement. The Jumbo Chrono boasts an exceptional 50mm size that is rarely seen in the world of luxury watches. The Jumbo Chrono's rhodium-plated rotor is unique and the impressive discovery dial reflects a strong graphic design: sweep-seconds chronograph in the center, hour and minute at 12:00, date at 2:00, 30-minute counter at 3:00, seconds at 6:00, and 12-hour counter at 9:00.

HUBLOT

BIG BANG BLACK FLUO – REF. 341.SV.9090.PR.0933

Measuring 41mm, and set with 198 black diamonds and 36 pink sapphires, this self-winding timepiece's black PVD-treated steel case sets a summery tone of vibrant opulence. Powering the watch with a frequency of 28,800 vph, the HUB4300 caliber animates a dial set with an additional 232 black diamonds. Read with tremendous legibility thanks to the thematic contrast between black and neon colors, the timepiece boasts a two-counter, central-seconds chronograph, along with a discreet date aperture at 4:30.

HUBLOT

BIG BANG FERRARI RED MAGIC CARBON – REF. 401.QX.0123.VR

Designed, developed, and produced entirely in house, this wristwatch's 330-component Unico movement with flyback chronograph boasts an extraordinary combination of reliability, robustness, and mechanical innovation. Visible on the dial side, the movement's famous column wheel serves as a window into the caliber's extraordinary construction, featuring a pallet fork and escapement wheel, both made from silicium and fixed to a removable platform. Framed by a 45.5mm case in carbon fiber and showcasing Ferrari's emblematic prancing horse at 9:00, an expressive red sapphire crystal dial is home to the automatic movement's flyback chronograph, with 60-minute counter at 3:00, within which a yellow date aperture is neatly positioned for optimal legibility.

HUBLOT

BIG BANG UNICO KING GOLD CERAMIC – REF. 411.OM.1180.RX

A 45.5mm 18K King Gold case sets a commanding tone for this self-winding timepiece with 72-hour power reserve. Driving the hours, minutes, small seconds at 9:00, date at 3:00, and flyback chronograph with 60-minute counter, the 330-component HUB 1242 Unico movement boasts a frequency of 28,800 vph and is entirely developed within the Hublot workshops. Complementing the timepiece's robust appearance with a sense of mechanical refinement, the caliber is revealed beneath SuperLumiNova-enhanced 5N gold-plated hands, indexes and counter rings on the matte black varnished skeleton dial.

HUBLOT

KING POWER BLACK MAMBA – REF. 748.CI.1119.PR.KOB13

Housed within a 48mm King Power case in micro-blasted ceramic, and brilliantly exposed through the sapphire crystal dial with anti-reflective treatment, the self-winding HUB 4248 caliber powers this 250-piece limited edition timepiece of decisive personality. Designed to commemorate the brand's partnership with its newest ambassador, basketball superstar Kobe Bryant, this 4Hz wristwatch boasts an inventive chronograph with peripheral 48-minute counter. Imbued with the Los Angeles Lakers' purple and gold theme, the King Power Black Mamba boasts a date window at 4:30, small seconds at 9:00 and an open micro-blasted ceramic caseback with Kobe Bryant signature.

IWC

INGENIEUR CHRONOGRAPH RACER – REF. IW378507

The Ingenieur Chronograph Racer animates its racing-inspired choreography on a slate-colored dial, within a 45mm case in stainless steel. At 12:00, beneath a red "60" that hints at the design codes of the MERCEDES AMG PETRONAS Formula One racecar, the flyback chronograph's red-tipped central seconds hand is complemented by an intuitive dual-function totalizer, while a peripheral tachometer scale permits the wearer to calculate the speed of a car over 1,000m. A subdial at 6:00 combines the small hacking seconds and rapid-advance digital date display.

IWC

INGENIEUR CHRONOGRAPH SILBERPFEIL – REF. IW378505

Named after the Silberpfeil (also known as the Silver Arrow), the dominant Mercedes-Benz racing cars of the 1930s, this self-winding stainless steel timepiece displays the hours, minutes, small seconds/date via a subdial at 6:00 and flyback chronograph with dual-function totalizer at 12:00. The silver-plated dial's intricate "perlage" décor pays tribute to the dashboard of the epoch's legendary Mercedes-Benz W25. The limited edition 45mm timepiece is engraved on its caseback with an elaborate depiction of a Silver Arrow racecar.

IWC

INGENIEUR DOUBLE CHRONOGRAPH TITANIUM – REF. IW386501

This 44mm titanium timepiece with silver-colored dial honors the high-speed racing world with a split-seconds chronograph that permits the measure of intermediate lap times without affecting the activity of the chronograph's primary hand. The day and date are told via two apertures at 3:00, while the racing-instrument theme of the watch is accentuated by two recessed chronograph totalizers at 12:00 and 6:00, and small seconds at 9:00. The self-winding timepiece is powered by the 79420 caliber and is worn on a black rubber strap with titanium pin buckle.

IWC

PORTUGUESE CHRONOGRAPH CLASSIC – REF. IW390402

Housed in an 18K red-gold case, the self-winding Portuguese Chronograph Classic honors the elegant cultural heritage of IWC's 1930s creations. The 42mm timepiece boasts a discreet date at 3:00, small seconds at 6:00 and a flyback chronograph whose slender blue central hand provides optimal contrast against the gold accents of the dial's Arabic numerals and hour and minute hands. A subdial at 12:00 combines the stopwatch's 12-hour and 60-minute totalizers into an easy-to-read display. Visible through a sapphire crystal caseback, the 89361 caliber, with 68-hour power reserve, is decorated with Côtes de Genève.

IWC

PORTUGUESE YACHT CHRONOGRAPH
EDITION "LAUREUS SPORT FOR GOOD FOUNDATION" – REF. IW390213

Standing out vibrantly against the depth and sheen of a dial in bright Laureus blue, a slender red hand indicates the elapsed seconds measured by the 89361 caliber's flyback chronograph. At 12:00, a dual-function totalizer indicates the stopwatch's 12-hour and 60-minute increments within an intuitive design. Completed with a date window at 3:00 and small seconds at 6:00, this stainless steel timepiece is engraved on its caseback with 12-year-old Hakkini Hasanga Sandumal De Silva's drawing, winner of IWC's annual "Time to move" competition, a reminder of the brand's devotion to helping needy children around the globe and of the motto of the Laureus Sport for Good Foundation.

LONGINES

THE LONGINES AVIGATION WATCH TYPE A-7 – REF. L2.799.4.53.0

The racy elegance of The Longines Avigation Watch Type A-7, a single pushbutton chronograph, immediately catches the eye. The steel case has an impressive diameter of 49mm and houses an L788 caliber, a column-wheel movement developed exclusively for Longines. The dial is angled at 50° to the right and features large white Arabic numerals, all of which make up the exceptional character of this timepiece. Its sleek appearance is enhanced by the tachometer scale around the outer edge of the dial and by the Breguet minute and hour hands. The whole is mounted on a black alligator strap. The intricate workings of this exceptional piece can be admired through the hinged cover on the sapphire crystal back.

LONGINES

THE LONGINES COLUMN-WHEEL SINGLE PUSH-PIECE CHRONOGRAPH REF. L2.774.8.23.X

The Longines Column-Wheel Single Push-Piece Chronograph models take their inspiration from the first chronograph wristwatches developed by Longines in 1913. With its 40mm diameter, this rose-gold model is fitted with the L788 caliber. This column-wheel single pushpiece chronograph movement has been developed by ETA exclusively for Longines. Its white dial is adorned with painted black numerals and a red "12," in a direct reference to the dials of the period. It is completed by two counters, at 3:00 and 9:00, together with a date aperture at 6:00. Blued hands harmoniously complete the whole.

PANERAI

LUMINOR 1950 3 DAYS CHRONO FLYBACK – 44MM – REF. PAM00524

Powered by an automatic-winding P.9100 caliber, executed entirely by Panerai, the Luminor 1950 3 Days Chrono Flyback – 44mm is complete with a power reserve of three days. The flyback chronograph function is controlled by a pushbutton at 8:00, which instantly returns the chronograph hands to zero and simultaneously restarts it without being necessary first to stop them and return them to zero by pressing the stop and reset buttons. The chronograph minutes and seconds hand are both positioned centrally, guaranteeing optimal legibility for the wearer.

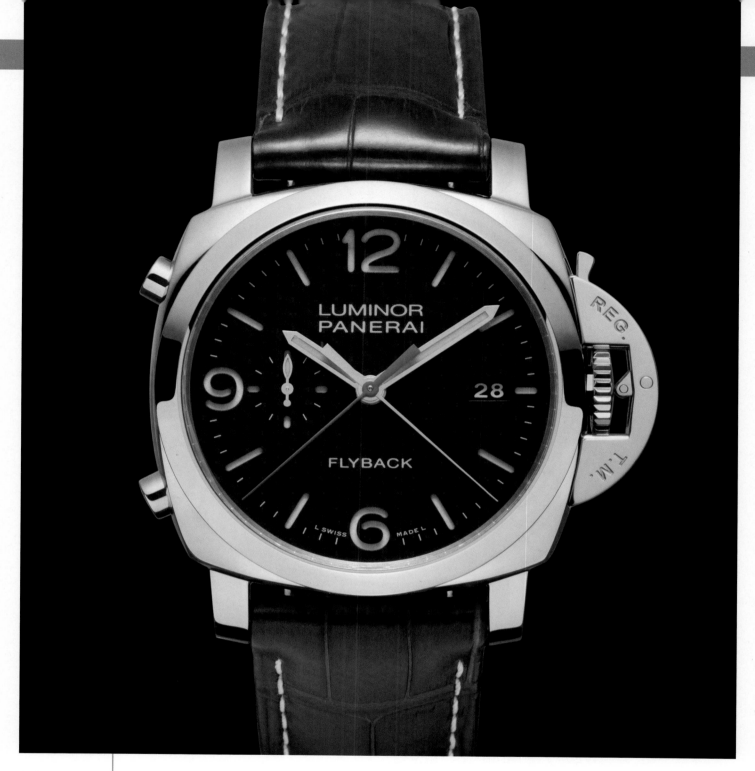

LUMINOR 1950 3 DAYS CHRONO FLYBACK ORO ROSSO – 44MM – REF. PAM00525

This new Panerai flyback chronograph is finished with an elegant, polished red-gold case. This 5Npt red gold is an alloy with a high percentage of copper, giving the color of the case great intensity, as well as a proportion of platinum, ultimately helping prevent oxidation. The signature Panerai bridge device which protects the crown is finished in the same material, ensuring a high level of water resistance for the timepiece. The gilted tone of these features highlights the luminous Arabic numerals and hour markers on the otherwise dark dial. The central hours and minutes are accompanied by a small seconds counter at 9:00, a date window at 3:00, and a flyback chronograph with a seconds reset function.

PANERAI

LUMINOR 1950 REGATTA 3 DAYS CHRONO FLYBACK TITANIO – 47MM – REF. PAM00526

The Luminor 1950 Regatta 3 Days Chrono Flyback Titanio marks the union of Panerai and the world of classic sailing. The simplicity of using the Regatta countdown function is evidence of the technical innovation achieved by the new P.9100/R automatic chronograph caliber, which is the heart and soul of this model. An orange pushbutton at 4:00 moves the central orange chronograph minute hand back one minute at a time until it is at the correct position in relation to the countdown. The wearer can start the chronograph by pushing the start/stop button at 10:00. Water resistant to 10atm, this model is truly the ultimate sailing piece.

PARMIGIANI

BUGATTI AEROLITHE – REF. PFP329-3400600

The Bugatti Aerolithe houses its self-winding PF 335 caliber in a racing-inspired 41mm case in titanium, embellishing its highly legible timekeeping display with the hypnotic depth of a dial in Abyss Blue. Accomplished through a state-of-the-art chemical process, the rich blue sheen of the timepiece is contrasted by the bright red accents of an ingenious 180° offset chronograph. Located on the left side of the case for thumb accessibility, the flyback chronograph's buttons permit the wearer to start, stop and reset the precise measure of time with a single action. Furthering the timepiece's devotion to optimal clarity, a large date aperture at 6:00 and small-seconds subdial at 9:00 complete the choreography of the 311-component caliber with two series-coupled barrels. The Bugatti Aerolithe is worn on an Hermès Epsom calfskin strap.

PATEK PHILIPPE

MEN COMPLICATIONS – REF. 5170G-001

This hand-wound wristwatch complements the timeless elegance of its white-gold case and hand-stitched alligator strap with the vintage aesthetic appeal of its silvery white dial's numerous railroad-track accents. Powered by the CH 29-535 PS caliber oscillating at a frequency of 28,800 vph, the 39mm timepiece boasts a column-wheel chronograph with 30-minute counter at 3:00. A subdial at 9:00 indicates the running seconds.

PATEK PHILIPPE

MEN NAUTILUS – REF. 5980/1AR-001

Upon a blue gradient dial that evokes the depth of color of a vast ocean, this self-winding steel and rose-gold timepiece complements its chronograph's central sweeping hand with an ingenious dual-function 12-hour/60-minute subdial at 6:00. With optimal clarity, the interior hand moves in half-hour intervals while the peripheral indicator is read accordingly as it conducts a full revolution every 30 minutes around its two-tiered scale. Powered by the 28,800 vph CH 28-520 C caliber, the timepiece is finished with a date window at 3:00, enhanced with luminescent gold hour markers and worn on a bracelet in accord with the case's two-tone structural theme.

PIAGET

PIAGET GOUVERNEUR CHRONOGRAPH – REF. G0A37112

The Piaget Gouverneur Chronograph houses the ultra-thin, self-winding 882P caliber, which is comprised of 33 jewels, possesses a power reserve of 50 hours and oscillates at 28,800 vph. Exhibiting technical expertise alongside a dashing appearance, this model is part of Piaget's Black Tie watch collection. The 18K pink-gold case matches the pink-gold appliqué hour markers, round counters and hands to a tee. These functions, which include a flyback chronograph as well as a second time zone counter and date window, rest beautifully atop a silvered guilloché dial.

PIAGET

PIAGET POLO FORTYFIVE CHRONOGRAPH – REF. G0A37004

A modern take on the renowned Piaget Polo watch, this handsome and athletic timepiece has a diameter of 45mm, which alludes to the 45-minute duration of a polo match. Piaget shows its daring side, as the case of this model is made of materials that were previously unused by the house. Composed of titanium and steel godroons with a black ADLC treatment, the exterior of this watch provides the wearer with maximum comfort. The black dial is complemented by bright white appliqué hour and minute markers as well as luminescent hands that guarantee legibility. Aside from the triple date window at 12:00 and the second time zone counter at 9:00, this watch possesses a flyback chronograph complete with a 30-minute counter at 3:00.

RICHARD MILLE

RM 50-01 TOURBILLON CHRONOGRAPH G-SENSOR LOTUS F1 TEAM ROMAIN GROSJEAN

Richard Mille Watches has been the official timekeeper of the Lotus F1 Team since 2013, and this fruitful collaboration is the first tourbillon watch dedicated to team member Romain Grosjean. RM 50-01 tourbillon uses a function indicator that shows whether the movement is in winding, neutral or hand-setting mode, with the addition of a mechanical G-force sensor, a mechanism able to transcribe the range of G's endured by the driver during deceleration phases. In total, the creation of this ultimate racing tool required more than 500 components.

RICHARD MILLE

RM 039 TOURBILLON FLYBACK AVIATION E6B

To explore the themes of aviation and aeronautics in greater depth, Richard Mille has created a caliber designed for flight navigation. This complete and technically sophisticated instrument is an ultra-modern interpretation of an essential device used by every established pilot. The watch displays much of the same information provided by the famous E6B flight computer. This round slide rule, incorporated into the bidirectional rotating bezel, can be used to read off and calculate fuel burn, flight times, ground speed, density altitude or wind correction, as well as the fast conversion of units of measurement. The RM 039 tourbillon caliber is also equipped with a UTC hand, a countdown mode, an oversize date at 12:00 and a function selector. The movement drives a flyback chronograph featuring an exclusive design.

ROGER DUBUIS

HOMMAGE CHRONOGRAPH – REF. RDDBHO0567

Roger Dubuis pursues its mission of reinterpreting the codes of classicism with a strikingly handsome chronograph that stands out boldly within a highly competitive field. The snailed chronograph counters at 3:00 and 9:00, complemented by a sweep seconds hand, bear eloquent testimony to the enduring strength of traditional watchmaking as applied to the modern task of accurate timekeeping. The combination of the concave bezel, high flange, applied five-minute markers, raised applied Roman numerals and subtly lowered center creates a mesmerizing impression of depth and multiple layers. The latter is in turn further emphasized by the audacious sunray guilloché dial motif, which itself echoes the radiating arrangement of the numerals and is picked up on the crown surface.

STÜHRLING ORIGINAL

AEVUS

The bright blue radiance of this 47mm wristwatch's blue PVD bezel endows the dial with a tone of dynamic masculine energy. Upon the dial, the contrasting hues of the indicators' blue, red and white accents accentuate an air of optimal legibility while displaying the ST-90888 quartz caliber's choreography of hours, minutes, central chronograph seconds, 30-minute and 12-hour counters, small seconds and a discreet quick-set date through a window at 4:00. A high-grade silicone rubber strap secures to the wrist this black PVD stainless steel watch, enhanced with generous luminous components and two skeletonized dauphine hands.

STÜHRLING ORIGINAL

PROMINENT

The textured ambience of this timepiece's gray guilloché-like stamped dial sets a captivating backdrop for a sophisticated display of timekeeping indications. Within the frame of a 40mm rose-toned stainless steel case, the self-winding ST-86912 caliber complements its 9:00 small seconds and 3:00 day and date apertures with a central-seconds chronograph complete with 30-minute and 12-hour totalizers at 12:00 and 6:00 respectively. Finished with eight Roman numerals applied on the inner circle, the Prominent is worn on an embossed brown alligator leather strap.

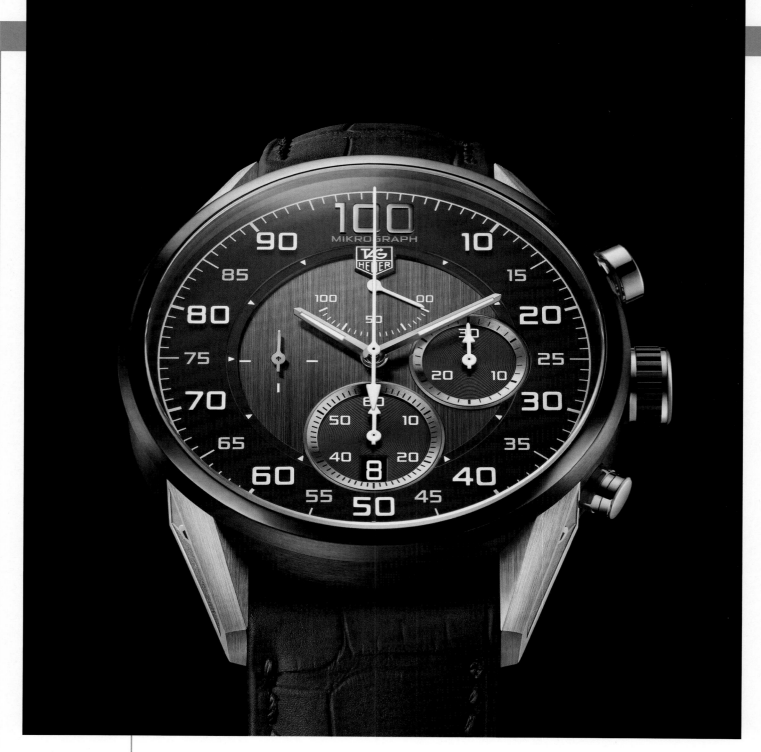

TAG HEUER

CARRERA MIKROGRAPH AVANT-GARDE – REF. CAR5A50.FC6319

Equipped with an easy-to-read 1/100th-of-a-second indication with central hand, this spectacular timepiece vibrates 360,000 times per hour. The chronograph's vibrations are unparalleled at 50Hz. This COSC-certified piece has a power reserve of 42 hours for the watch and 90 minutes for the chronograph. The design is spectacular, with a 5N 18K solid rose-gold and titanium case coated with titanium carbide. The dial is equally stunning with a black exterior and fine-brushed anthracite interior. This magnificent piece sits on a high-tech soft-touch black alligator leather strap.

TAG HEUER

CARRERA MIKROPENDULUM – REF. CAR2B83.FC6339

Like all MIKRO creations, the TAG Heuer Carrera Mikropendulum contains a dual chain platform with a balance-wheel system for the watch. Oscillating at 28,800 vph with a 42-hour power reserve, this model is unique in that it operates via a hairspring-less pendulum system for the chronograph. This piece is based on the Pendulum's technology and is the world's first regulator to work with magnets, overturning three centuries of conventional watchmaking. The effect of gravity on a classical hairspring is a pressing issue, as its shape gets slightly altered, which can generate errors in time measurement. Thanks to the Pendulum, this problem is not present in the Carrera Mikropendulum.

VACHERON CONSTANTIN

OVERSEAS CHRONOGRAPH – REF. 49150/B01A-9745

Beating with a frequency of 21,600 vph, the Overseas Chronograph's automatic-winding 1137 caliber provides this timepiece with a precise set of functionalities. At 3:00 and 9:00, respectively, the chronograph's 30-minute and 12-hour counters join the small seconds at 6:00 and large date at 12:00, to complete the backdrop of the frontal view. While the stainless steel case and blue dial give the watch a powerful allure, the adornment of the caseback with the two-masted Overseas ship medallion reminds the wearer of watchmaking's rich nautical heritage.

ZENITH

EL PRIMERO LIGHTWEIGHT – REF. 10.2260.4052W/98.R573

This 45mm avant-garde masterpiece rejects the boundaries of conventional watchmaking with a groundbreaking movement weighing a mere 15.45g in spite of its 334 components. Within a carbon case with inner structure in ultra-light and ultra-hard ceramised aluminum, the self-winding titanium El Primero 4052 caliber boasts a flyback chronograph whose red central hand completes a revolution of the skeletonized dial every ten seconds. At 6:00, a disc-driven date function confirms the cutting-edge aesthetics of the timepiece, while the three-color, three-counter, central display pays homage to the El Primero's iconic codes of design.

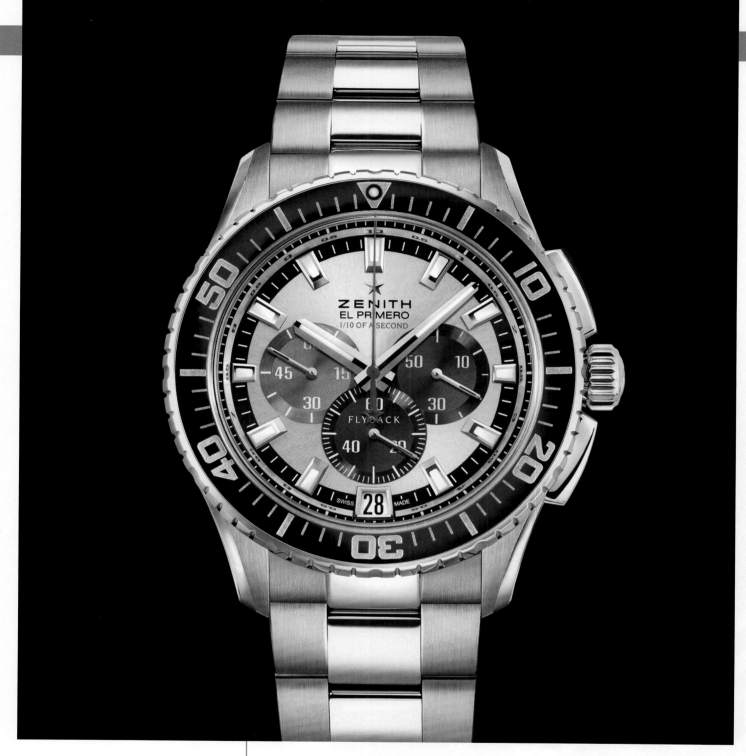

ZENITH

EL PRIMERO STRATOS FLYBACK STRIKING 10TH – REF. 03.2062.4057/69.M2060

First watch to break the speed of sound in a near space environment with Felix Baumgartner, the El Primero Stratos Flyback Striking 10th is equipped with the world's most accurate automatic chronograph movement: the legendary El Primero. Vibrating at a rate of 36,000 vph, El Primero Calibre 4057 B is framed by a generous 45.5mm case that is water resistant to 100m. A worthy descendant of the Rainbow Flyback model developed for the French Air Force in 1997, this impressive chronograph boasts a flyback function in addition to its capacity to display 1/10ths of a second.

Alarms

Vulcain

Revolution GMT Automatic

Alarms

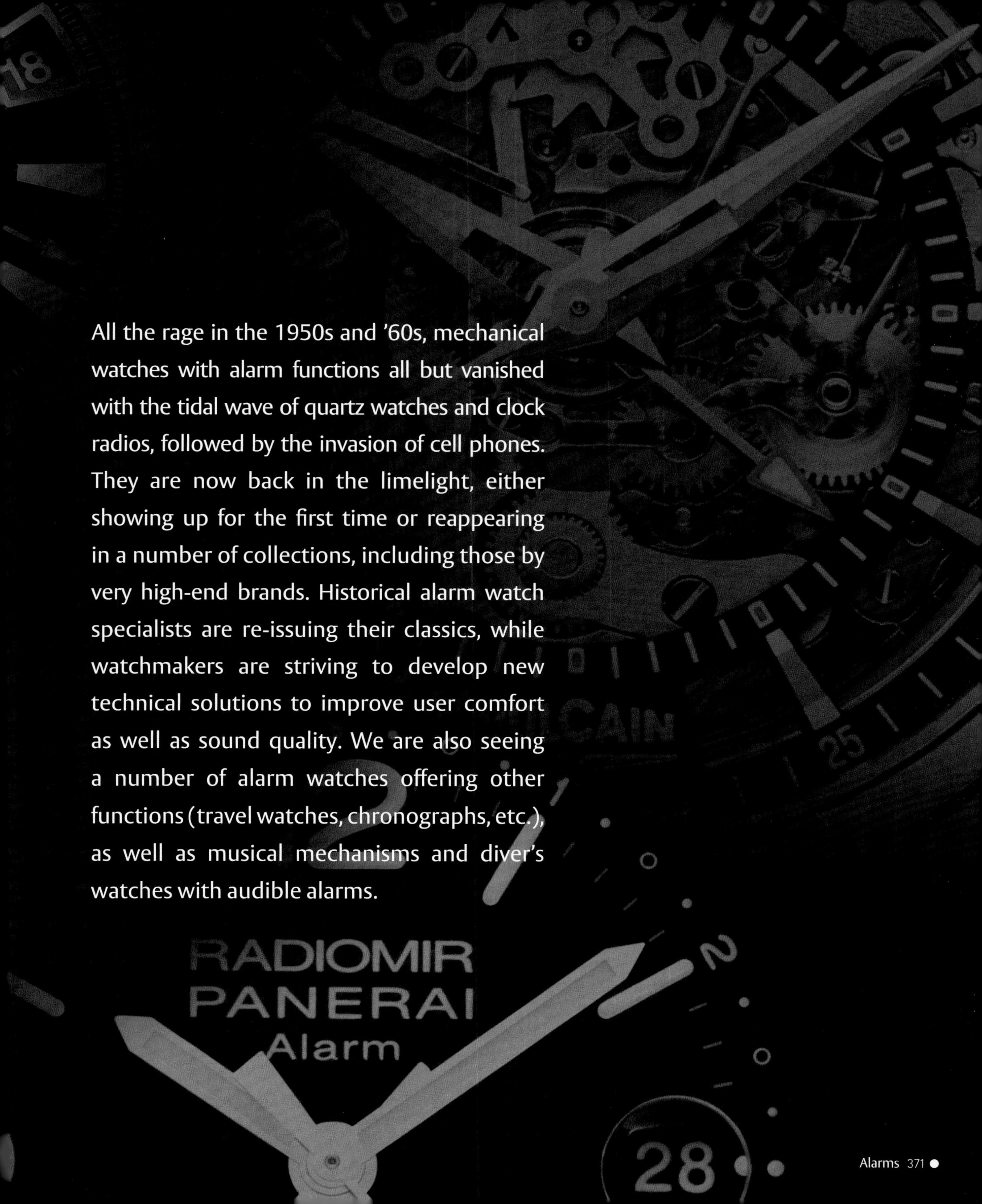

All the rage in the 1950s and '60s, mechanical watches with alarm functions all but vanished with the tidal wave of quartz watches and clock radios, followed by the invasion of cell phones. They are now back in the limelight, either showing up for the first time or reappearing in a number of collections, including those by very high-end brands. Historical alarm watch specialists are re-issuing their classics, while watchmakers are striving to develop new technical solutions to improve user comfort as well as sound quality. We are also seeing a number of alarm watches offering other functions (travel watches, chronographs, etc.), as well as musical mechanisms and diver's watches with audible alarms.

RADIOMIR
PANERAI
Alarm

THE ALARM WATCH IS UNDOUBTEDLY THE OLDEST HOROLOGICAL COMPLICATION.

In fact, it seems that in the 13th century, Western monasteries were equipped with audible mechanisms intended to awaken monks so they could say morning prayers at the prescribed hour. One should also note that the word "watch," which comes from the same root as "wake," first referred to "a clock that wakes one up," before being applied to any watch, no matter what functions it provided.

In the 15th century, table clocks were often equipped with an alarm mechanism that sounded on a small bell. This type of instrument would enjoy such high consideration that in 1601, the rules of the Corporation des Horlogers de Genève would demand that whoever wished to earn the certification of "maître" or master-watchmaker must produce a small clock with a morning alarm. The 16th century saw the appearance of pocket alarm watches with openworked casebacks to enhance the acoustics of the sound. In the 17th century, watchmakers also created oversized "carriage watches," with mechanisms to wake passengers in time to make their connections.

The first wristwatch alarms began to appear in 1914, but their sound was not yet strong enough to wake a sound sleeper. For that, the world would have to wait until 1947, when Vulcain launched the first truly functional alarm wristwatch, powered by the famously piercing Cricket caliber.

In the 1950s and '60s, the mechanical alarm watch experienced its golden age. Many watch brands were interested in this newly popular domain, developing many technical innovations, such as the first watches with a volume option (loud/soft) or with self-winding movements.

However, in the 1970s, the tidal wave of quartz movements and radio alarm clocks seemed to sign the death warrant for mechanical alarm watches.

The renaissance of mechanical watchmaking, at the end of the 1980s, signaled the grand return of the wristwatch alarm. For several years, this grand classic of horology has been enjoying renewed popularity among some of the biggest names in the industry. Technical enhancements abound, as do variations on display systems and designs—all of which are clear signs that this useful function will be making itself heard for some time to come.

The alarm watch mechanism comprises two parts, one for setting the time, the other to activate the alarm strike. As a rule, it uses its own barrel (energy storage), separate from the barrel that powers the movement.

Alarm wristwatches most often distinguish themselves by the presence of an additional crown, which allows the wearer to both reset the spring of the striking mechanism and set the alarm.

Certain models also boast a small display on the dial indicating if the alarm is engaged or not.

The sound is produced by a small hammer that strikes a piece of metal. This metal piece could be a membrane with a pin, a kind of bell set into the case middle, or a gong (a steel wire wrapped around the movement), as in minute repeaters. Gongs produce a less powerful sound than the other two solutions. The alarm sounds until the striking work barrel is completely wound down.

Rereleasing the Classics

The history of the alarm watch has its stars, its legends, its flagship models. Riding the wave of enthusiasm for all things vintage, several famous brands have rereleased certain classic pieces, adapting them to modern standards of size and technical sophistication.

Jaeger-LeCoultre
Master Memovox

In 1956, Jaeger-LeCoultre launched the first automatic watch with an alarm function. The manufacture paid homage to this iconic model by creating the Master Memovox, with caliber 956 integrating the latest technical developments of the manufacture.

Tudor, *Heritage Advisor*

Inspired by a 1957 model, the Heritage Advisor from Tudor is equipped with an alarm module developed by the brand. The ON/OFF window at 9 o'clock indicates if the alarm function is engaged and the alarm's power reserve is displayed on a disc at 3 o'clock.

Vulcain, *50s Presidents' Watch*

Tribute to the famous "Presidents' Watch," first released in 1947, the 50s Presidents' Watch from Vulcain combines a design inspired by a 1950s model with the modern performance of the automatic-winding alarm caliber Cricket V-21.

Corum, *Heritage Vintage Chargé d'Affaires*

A reinterpretation of a model first launched in 1956 in the first Corum collection, the Heritage Vintage Chargé d'Affaires limited series incorporates the same alarm movement as the original and rings for a full 12 seconds.

Jaeger-LeCoultre, *Master Memovox International*

Jaeger-LeCoultre celebrated the 1958 launch of the Memovox Worldtimer (also known as Memovox "Heures Internationales") by creating the Master Memovox International. The dial's central disc, which allows the wearer to set the alarm time, is adorned with the names of 24 cities, representing the 24 time zones.

Variations on a Theme

In its most classic version, the alarm watch stands out by its small additional hand or the central disc for regulating the alarm. But the choice is ever-expanding, with multiple variations on the display mode or the style—whose openworked dials reveal the mechanisms within. On the technical side, the alarm increasingly accompanies other useful complications, such as the chronograph or the calendar.

⋀ Alfex, *Pazzola Giant Alarm*

In 2009, the brand from Tessin, Switzerland, Alfex, celebrated its 60th anniversary by presenting the Pazzola Giant Alarm. One technical and aesthetic innovation: setting the alarm is done via a disc placed under the dial, positioned in relation to a red marker at 6 o'clock.

⋖ Oris, *Artelier Alarm*

Oris—known for its small alarm clocks in the 1940s—now offers an elegant, classic wristwatch named Artelier Alarm, with a central display separate from the alarm time and a chime specially developed by the brand.

⋁ Fortis

B-43 Flieger Chronograph Alarm GMT

Created for the 100th anniversary of Fortis, the B-43 Flieger Chronograph Alarm GMT boasts the first automatic chronograph movement with integrated alarm function—completed by a second time zone.

⋀ Vulcain, *Golden Heart*

Due to its partially skeletonized dial, Vulcain's Golden Heart allows the wearer to admire the complex architecture of the manufacture's Cricket movement, with its anthracite coating and carefully executed finishings.

Jaeger-LeCoultre, *Master Grand Réveil*

A rare, even unique, example of an alarm watch with perpetual calendar, the Master Grand Réveil from Jaeger-LeCoultre can be set to either chime or vibrate. The alarm mechanism includes a barrel independent from the principal movement, as well as a hammer that makes a gong suspended in the caseback vibrate.

Jaeger-LeCoultre, *AMVOX1 R-Alarm*

Born from the partnership between Jaeger-LeCoultre and Aston Martin, the AMVOX1 R-Alarm stands out for its sporty aesthetic that recalls automobile gauges. The alarm is set with a small hand on a minute track at the center of the dial.

Alarms one Can Set in Advance

On a traditional alarm watch, the alarm can only be set 12 hours in advance, i.e. in the evening for the next morning. For increased practicality, several brands have developed mechanisms that can set the alarm 24 hours ahead of time—or even a month in advance. Certain systems also set the alarm to the minute.

Harry Winston, *Project Z6 Black Edition*

Harry Winston has put its name to a highly original reinterpretation of the alarm watch in its limited edition Project Z6 Black Edition. The small off-center dial at 4 o'clock allows the wearer to set the hour and minute of the alarm up to 24 hours in advance, using two discs and a day/night indicator. The rectangular gong, cast in one piece, guarantees optimal acoustics.

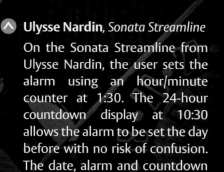

Ulysse Nardin, *Sonata Streamline*

On the Sonata Streamline from Ulysse Nardin, the user sets the alarm using an hour/minute counter at 1:30. The 24-hour countdown display at 10:30 allows the alarm to be set the day before with no risk of confusion. The date, alarm and countdown automatically adjust to the selection of a new time zone.

Glashütte Original
Senator Diary

The German brand Glashütte Original, a branch of the Swatch Group, innovated with its presentation of the Senator Diary, the first alarm watch programmable for up to 30 days in advance. The crown at 10 o'clock and the pushbutton at 8 o'clock allow the wearer to determine the date and time of the alarm to within a quarter of an hour.

Musical Alarms

Moving away from more strident sounds, certain brands prefer to charm the ear by equipping their alarm watches with "cathedral" chimes worthy of the most beautiful minute repeaters, or even the kind of precious mechanisms that play entire tunes, as do music boxes. These dulcet tones soften the harsh reality of waking up—or remind the wearer with a few notes of an important meeting during the day.

Vulcain, *Imperial Gong*

The first wristwatch with an alarm and a tourbillon, Vulcain's Imperial Gong—launched in 2005—also stands out for its cathedral chime, worthy of the most beautiful minute repeaters.

Pierre DeRoche, *TNT RendezVous*

The TNT RendezVous from Pierre DeRoche is not designed to wake a slumberer. It prefers to discreetly signal a moment chosen in advance by a series of double chimes. Its two hammers strike two "cathedral" gongs partially visible through the dial.

Boegli, *Grand Orchestre*

Equipped with a tiny 17-note music box, the Grand Orchestre model by Boegli offers a chance to wake up gently to the sound of Mozart's "Magic Flute," Vivaldi's "Four Seasons," or Beethoven's "Für Elise."

Breguet, *Réveil Musical*

Masterpiece of haute horology, Breguet's Réveil Musical possesses a patented mechanism inspired by music boxes, with the cylinder replaced by a disc with pins. During the melody, which lasts about 20 seconds, the dial completes a full rotation. The automatic caliber is equipped with a silicon escapement.

Travel Alarms

What is more useful than an alarm while one is traveling? Inspired by this rhetorical question, watchmakers are increasingly combining the alarm function with a second time zone or "world time" display (see chapter on Multiple time zones and GMTs). The most sophisticated models stand out due to their alarm mechanism that is always linked to the local time.

Panerai, *Radiomir GMT Alarm*

The Radiomir GMT Alarm from Panerai combines a digital display of a second time zone at 6 o'clock with an alarm indicated by a central hand. The two functions are controlled by a crown located on the right side of the case.

Blancpain, *Léman GMT Alarm*

On the Léman GMT Alarm from Blancpain, the alarm is indexed to the local time display, indicated by the large hour and minute hands. Another practical innovation: the automatic-winding mechanism allows the wearer to charge the power reserve and the alarm simultaneously.

Vulcain, *Aviator GMT Pilot DLC*

The Aviator GMT Pilot DLC model by Vulcain combines a hand-wound alarm caliber with a world-time display. The inner bezel ring indicating the 24 reference cities is adjusted via the crown at 4 o'clock.

Breguet, *Le Réveil du Tsar*

The Réveil du Tsar, from Breguet, possesses an alarm mechanism indexed to the local time, with an alarm engagement indicator in a small window at 12 o'clock. The hours/minutes subdial, at 3 o'clock, allows the wearer to set the alarm with to-the-minute precision.

Jaeger-LeCoultre
Master Compressor Extreme W-Alarm Aston Martin, Limited Edition

The Master Compressor Extreme W-Alarm Aston Martin from Jaeger-LeCoultre, with a world time function, presents several patented innovations in its alarm mechanism, which is equipped with a gong wrapped around the movement. The time of the alarm can be precisely selected using a double disc with hours and minutes in a window at 9 o'clock.

Yeslam, *Ti2 Réveil*

Equipped with a movement from La Joux-Perret, the Ti2 Réveil from Yeslam contains within its square titanium case an alarm with a central hand, an aperture for the date and a 24-hour time zone with a digital display.

Underwater Sounds

An ever-increasing number of diving watches are equipped with an alarm, and more and more alarm watches are getting their sea legs. This might seem like an odd alliance, as the ocean depths are not ideal for spending the night, or even for a small nap. But here the alarm is used to remind the wearer of a precise moment, notably the end of the dive time—and it doesn't hurt that sound carries much better underwater than in open air.

> ### Vulcain
> *Nautical Heritage*

In a rerelease of a famous 1961 model, the Nautical Heritage from Vulcain—water resistant to 300m—stands out for its alarm that is audible underwater, its central rotating dial displaying decompression stages and its triple-backed case, which serves as a resonance chamber for the chime.

> ### Jaeger-LeCoultre
> *Master Compressor Diving Alarm Navy SEALs*

Jaeger-LeCoultre celebrated its partnership with the elite combat swimmers through the 2009 launch of the Master Compressor Diving Alarm Navy SEALs, a limited edition with a grade-5 titanium case, a ceramic bezel and water resistance to 300m.

> **Breguet**, *Marine Royale*

Breguet has released its sporty Marine Royale collection in an alarm model water resistant to 300m, with a modern aesthetic that combines rubber and rose gold.

Vulcain, *Diver X-Treme Automatic*

Vulcain has developed several watches that combine an alarm with underwater capability, such as the Diver X-Treme Automatic, with a case that is water resistant to 100m, audible underwater strike, rotating 60-minute bezel to calculate the dive time and an ultra-readable dial highlighted with luminescent indexes.

Jaeger-LeCoultre, *Memovox Tribute to Deep Sea*

In homage to Memovox Deep Sea, the first Jaeger-LeCoultre diving watch, released in 1959, the Memovox Tribute to Deep Sea is powered by the automatic Memovox 956 caliber, with hours, minutes, seconds and alarm. The caseback reprises the timepiece's original design of a frogman surrounded by bubbles.

BREGUET

CLASSIQUE LA MUSICALE – REF. 7800BA.11.9YV

A masterful fusion of innovation, sublime craftsmanship, and design ingenuity, this self-winding 48mm timepiece in 18K yellow gold conducts a brilliant visual accompaniment to its pioneering musical score. Equipped with the first ever magnetic strike governor, a patented system of isolated magnets used to eliminate background noise, the melodic tune, pre-set to a desired time or activated via a pushbutton at 10:00, is joined by the superb and complete rotation of the hand-guilloché dial. Powered by the 55-jewel 0900 caliber with silicon escape wheel and Breguet balance and spring, this horological work of art is finished with a caseband intricately engraved with a musical stave.

VULCAIN

50S PRESIDENTS' WATCH – CRICKET PRESIDENT "EDITION FRANCE" – REF. 160151.323L

Paying homage to France and to the enthusiasm for the brand shown by French customers, Vulcain is innovating by introducing its first ever watch specifically dedicated to a nation. This 50-piece limited edition of the iconic 50s Presidents' Watch features a 42mm steel case housing the legendary mechanical hand-wound Cricket V-16 alarm movement equipped with the brand-patented Exactomatic system ensuring improved rating regularity on the wrist. Particular care has been lavished on the polished and satin-brushed finish of the case that lends a chic and vintage touch. The sunburst silver-toned dial with applied hour markers is distinguished by its aura of elegant understatement expressed through hours, minutes and seconds hands along with a date window at 6:00.

VULCAIN

50S PRESIDENTS' WATCH – AUTOMATIC – REF. 210550.279L

A Fifties spirit expressed through a contemporary model: such was the intention behind the creation of the new 50s Presidents' Watch collection by Vulcain. This alarm watch is inspired by a model from the Fifties, while meeting the highest modern standards. Powered by the Cricket V-21 alarm self-wound caliber, the 50s Presidents' Watch is comprised of 28 jewels and 263 components. Oscillating at 18,000 vph, the Cricket V-21 caliber is decorated with rhodium coating and Côtes de Genève motif. Housed in a 5N 18K pink-gold case, this piece is water resistant to 5atm.

ZENITH

PILOT DOUBLEMATIC – REF. 03.2400.4046/21.C721

The Pilot Doublematic is a open invitation to travel. This chronograph exercises infallible control over time and features an alarm function controlled by the pushbutton and the crown at 8:00. The Pilot Doublematic also defies time differences thanks to its two discs simultaneously displaying the time in different parts of the world. Beating at 36,000 vph and endowed with a 50-hour power reserve, the mechanical automatic-winding El Primero Calibre 4046 may be admired through an openworked oscillating weight that is itself visible through the sapphire caseback.

Multiple Time Zones and GMTs

de Grisogono

Instrumento

Multiple Time Zones

MT

One hundred and thirty years after the Earth was divided into 24 time zones (in 1884), globalization has become the name of the game. Today is all about travel and telecommunications around the planet. To meet this ever-growing need to juggle time zones, watchmakers are constantly developing new watches that provide simultaneous displays of the time in several places around the world, while focusing firmly on user-friendliness and readability. The result is that among all the functions provided by a timepiece, this horological complication is doubtlessly the one that now appears in the most varied guises—ranging from a simple double-handed watch to the famous world-time or GMT watches, and including multiple systems involving disks, rotating bezels, planispheres and even tiny 3D globes.

BEFORE THE DIVISION OF THE GLOBE INTO 24 TIME ZONES, EACH REGION LIVED AT ITS OWN PACE, AND TRAVELERS HAD TO ADJUST THEIR WATCHES CONSTANTLY AS THEY MOVED FROM ONE PLACE TO ANOTHER.

There were a multitude of local times, based, more or less, on longitude: more than 70 in North America and around 30 in Europe. The increase in long-distance travel, and especially the birth of the railroad, generated a need to create a unified system of time. In 1847, the British Railway Clearing House recommended that all companies adopt the time at the Greenwich Observatory (Greenwich Mean Time, or GMT), which already served as a reference for navigators. In 1884, at the International Meridian Conference in Washington, DC, the decision was taken to divide the planet into 24 time zones, each one covering 15° of longitude (one hour), and to assign longitude 0 to the Greenwich meridian. This system was rapidly adopted by most of the countries in the world—except France, which did not come around until 1911.

The first dual time zone watches, sometimes called "captain's watches," were equipped with two movements, more or less coordinated, that displayed the time zones on two different dials. Later, watchmakers perfected this system by using just one movement, equipped with a mechanism uncoupling the additional hour hand.

Though watches with a multitude of local times already existed as early as the 17th century, it was not until the 1930s that a Genevan watchmaker, Louis Cottier, invented the ingenious "world time" system, which allows the wearer to read the time in all time zones simultaneously.

GMT AND UTC

Greenwich Mean Time corresponds to the mean solar time at the meridian crossing the Greenwich Royal Observatory, near London. It was adopted at the end of the 18th century as a reference time zone for all others on the planet, and is based on the rotation of the Earth. It was replaced in 1972 by "Universal Coordinated Time," better known by its acronym UTC, which is based on International Atomic Time (with the acronym TAI, for its French name). Watchmakers commonly use the term GMT to designate a watch indicating a second time zone, even if that time zone is not Greenwich.

HOW DOES IT WORK?

Multiple time zone watches offer such a variety of technical approaches and methods of use, that it is impossible to describe in detail all the particularities of their functioning.

Generally, they possess an additional display for the hour that is powered by the main system, but can be temporarily disengaged in order to set it to local time—without affecting the accuracy of the minutes and seconds. A pushbutton or the crown serves as a tool for setting the second time zone.

In order not to awaken a friend or a business partner who might be sound asleep, travelers often need to know if it is day or night in their home region. This is why some watches display home time on a 24-hour scale. Others use a 12-hour display, but supplement this with a day/night indication (a small aperture or subdial that changes from white to black according to the time of day), a small additional 24-hour scale or an AM/PM display.

Many watches boast another very useful function when one is traveling: automatic linkage of the date to local time.

WORLD TIME

To continuously display the time in all corners of the globe, world-time watches possess—in addition to the central hands—two mobile discs, one with 24 gradations, the other bearing the names of the reference cities. The 24-hour disc turns counterclockwise.

Double Pointer-Type Systems

The simplest system for simultaneously displaying two time zones is to add an hour hand with a distinctive design, either to the center of the dial, or—more rarely—on a small auxiliary dial. The second time zone might also distinguish itself with a retrograde or jumping display. The most practical models equip their home time zone with a 24-hour scale or a day/night indicator. Luxury watchmakers also take great care to make the systems for setting the time as easy to use as possible.

> **Zenith**, *Montre d'Aéronef Type 20 GMT*
>
> Inspired by the vintage Zenith aviator watches, the Montre d'Aéronef Type 20 GMT, from the Pilot collection, indicates a second time zone in 24-hour mode by means of a hollowed red-tipped hand.

> **Eterna**, *Royal KonTiki Two Time Zones*
>
> On the Royal KonTiki Two Time Zones from Eterna, equipped with a manufacture movement, the red-tipped second time-zone hand runs over a light and dark-shaded 24-hour scale indicating daytime and night-time hours.

IWC, *Ingenieur Dual Time Titanium*

On the Ingenieur Dual Time Titanium by IWC, the main hands show the local time, while the hand with the white triangle provides a 24-hour dual-time display. Pushpieces serve to adjust the time in one-hour increments.

DeWitt, *Academia Double Fuseau GMT II Poetic*

The Academia Double Fuseau GMT II Poetic from DeWitt offers an original GMT complication. The wearer sets the second time zone, indicated by a red central hand, by pinpointing its position on the tiny globe at 12 o'clock.

Rolex, *Oyster Perpetual GMT-Master II*

The Rolex Oyster Perpetual GMT-Master II is a classic dual time-zone watch launched in 1955 that solves the problem of daytime and nighttime hours by means of a 24-hour display of the second time zone along with a two-tone blue and black ceramic bezel.

Louis Vuitton
Tambour éVolution GMT

The second red-tipped hand of the Tambour éVolution GMT, by Louis Vuitton, completes a revolution in 12 hours, but is complemented by a day/night indicator for the second time zone.

de GRISOGONO, *Fuso Quadrato*

On the Fuso Quadrato from de GRISOGONO, the hand for the additional time zone is hidden beneath a dial composed of 12 titanium shutters that opens on demand using a slide at 9 o'clock.

◀ **Cartier**
Ballon Bleu de Cartier
Flying Tourbillon 2nd time zone

The Ballon Bleu de Cartier Flying Tourbillon 2nd time zone watch is inspired by regulator-type dials with a central minute hand and two jumping-hour totalizers at 10 o'clock and 2 o'clock respectively.

Superimposable Hands

On most travel watches, the additional time zone hand is always visible, even when the wearer has no use for it. Some watchmakers have improved upon this "defect" by devising hour hands that can be superposed, or placed one on top of the other, and function in a perfectly synchronized manner. This is a means of offering two watches in one.

ᴧ **Patek Philippe**
Aquanaut Travel Time Ref. 5164
On the Aquanaut Travel Time from Patek Philippe, the pushbuttons allow the wearer to bring the local hour hand forward or back in increments of one hour. When the dual time zone function is not in use, the two hour hands appear as one.

ᴧ **Jaquet Droz**, *Grande Heure GMT*
Bearing just two superimposed compass-type hour hands—a nod to historical navigating instruments—the Grande Heure GMT by Jaquet Droz provides an intuitive 24-hour dual-time reading.

ᴧ **Jaeger-LeCoultre**, *Master Hometime*
The Master Hometime from Jaeger-LeCoultre proposes a system of dual crown-activated superimposable hour hands. The skeletonized hand on a 24-hour scale maintains home time. The date is linked to the local time.

❯ **H. Moser & Cie**, *Nomad Dual Time*
On the self-winding Nomad Dual Time by H. Moser & Cie, two superimposed hour hands are set via a patented crown-adjustment system. The white or black indicator at 12 o'clock distinguishes between day and night (AM or PM hours) in the wearer's home time zone.

Three-Dimensional Planets

After the appearance of three-dimensional moons (see chapter on Moonphases), a new type of travel watch emerged, one that displayed small terrestrial globes rotating on the dial. These innovative, spectacular creations are most often produced by young, pioneering brands or independent watchmakers.

> **Greubel Forsey**, *GMT*

The titanium globe adorning Greubel Forsey's GMT allows the wearer to take in at a glance all the Earth's time zones—complemented by a day/night indicator. For a more precise reading, the watch also displays a second time zone at 10 o'clock and a world time indication on the caseback.

> **Hysek**, *Colosso*

On the Colosso from Hysek, the terrestrial sphere, rotating over 24 hours, can be linked to either local time (on the central hands) or to a second time zone. The latter is indicated by a double linear retrograde display, with an aperture that specifies the reference city.

> **Magellan Watch**
> *Magellan 1521 NH*

Using a third hand tipped with a small solar disc, the Magellan 1521 NH displays either the position of the sun in relation to the Earth (longitude where it is noon), local time on a 24-hour scale or a second time zone, also on a 24-hour scale.

> **VicenTerra**, *GMT-3*

On the VicenTerra GMT-3, the small rotating sphere placed at 5 o'clock shows the longitude where the sun is at its height. This unusual system is completed by a second time zone at 7 o'clock and a day/night indication (sun and moon) at 12 o'clock.

Double Hour/Minute Displays

Though the minutes generally remain the same from one time zone to the next, certain high-end watch brands prefer to include a complete display of the time in each region, with a second set of both hour and minute hands located in a small auxiliary dial. Another possibility involves indicating the two time zones on two separate dials, in watches featuring a resolutely distinctive design.

Hermès, *Cape Cod GMT*

The Cape Cod GMT by Hermès provides a second time zone in hours and minutes on a subdial at 6 o'clock, along with a two-tone day/night indication. It is equipped with a self-winding movement visible through the transparent caseback.

Piaget, *Altiplano Double Jeu*

Piaget's Altiplano Double Jeu possesses two superimposed cases, each containing an ultra-flat mechanical movement. The dial of the lower case displays a second time zone on a 24-hour scale.

Icelink, *Ambassador*

The brand Icelink boldly opts for an extra-extra-large size in its Ambassador line—the better to fit six quartz movements that display the time in six cities around the world.

Breguet, *7067 Tradition GMT*

Equipped with an openworked architecture that reveals the principal components of the movement, the 7067 Tradition GMT from Breguet displays local time at 12 o'clock. A guilloché subdial at 8 o'clock displays home time, complete with day/night indicator.

Jaeger-LeCoultre, *Reverso Grande GMT*

The Reverso Grande GMT from Jaeger-LeCoultre makes the most of its famous reversible dial by proposing two time zones on two dials that lie back to back. The dial on the reverse side includes an indicator for the number of hours' time difference between local time and Greenwich Mean Time or another reference point.

de GRISOGONO, *Meccanico dG*

de GRISOGONO's astonishing Meccanico dG combines an analog display for the first time zone and a digital display, driven by 23 cams, for the second time zone.

Adjustment by Half Time Zones, Quarter Time Zones, or to the Nearest Minute

On the vast majority of travel watches, the second time zone is adjustable in increments of one hour. But there are countries in which the deviation from UTC adopts finer divisions of half an hour or even a quarter of an hour. This is notably the case with Iran (UTC + 3h30), India (UTC + 5h30) Nepal (UTC + 5h45) and the middle of Australia (UTC + 9h30). A few recent watches take these particularities into account.

∧ Glashütte Original
Grande Cosmopolite Tourbillon

The Grande Cosmopolite Tourbillon from Glashütte Original can display two hour/minute time zones from among a selection of 37 locations (including those that vary by 30 or 45 minutes), with a distinction between summer and winter times. Thanks to its perpetual calendar, it also takes into account the exact monthly and leap-year cycle for these calculations.

∧ Blancpain, *Villeret 8 Day Half-Timezone*

The Villeret Timezone 8 Day Half-Timezone from Blancpain allows the wearer to set the second time zone (which includes hours and minutes) in increments of half an hour using a crown equipped with an integrated pushbutton.

< Seiko, *Astron GPS Solaire*

Thanks to its integrated GPS receiver, the Astron GPS Solar by Seiko is capable of recognizing all 39 time zones in use around the world. The hands are thus guided in such a way as to automatically adjust to local time with atomic-clock accuracy.

> **Arnold & Sons**, *DBG*

The DBG by Arnold & Son features two separate hour/minute displays, powered by their own barrels, going trains, escapements and balances, and independently adjustable to the nearest minute. The subdial at 12 o'clock displays the difference between the two time zones.

> **Jaeger-LeCoultre**
> *Duomètre Unique Travel Time*

The Duomètre Unique Travel Time from Jaeger-LeCoultre displays a second time zone on a subdial at 9 o'clock with a jumping hour and minute hand adjustable to the nearest minute. The globe at 6 o'clock enables an intuitive reading of the time around the world.

Systems with Discs or Rotating Bezels

Watches with multiple time zones might operate using systems based on discs or rotating bezels—either in lieu of a pointer-type system or in conjunction with it. Watchmakers generally take this opportunity to introduce complementary indications, such as the names or abbreviations for certain cities, representing precise time zones.

> **Cartier**, *Tortue XXL Multiple Time Zone*
> On the Tortue XXL Multiple Time Zone watch by Cartier, a pushpiece and a lateral city disc serve to select a second time zone while taking account of "summer" or daylight saving time. The local time is displayed by the main hands, while the time in the user's home zone appears through an aperture with a day/night hand incorporating symbols of the sun and moon.

Jacob & Co., *Epic SF 24*

The Epic SF 24 model by Jacob & Co. shows the time in one of the world's 24 time zones through a large aperture at 12 o'clock, in a style inspired by airport display boards.

Ralph Lauren, *Sporting World Time*

The Sporting World Time by Ralph Lauren, equipped with a Jaeger-LeCoultre movement, provides an hour/minute second time-zone display on a subdial at 6 o'clock, complete with a day/night indicator. The city corresponding to this second time zone may be selected by rotating the crown at 10 o'clock.

Montblanc
Nicolas Rieussec Chronograph Open Home Time

The Nicolas Rieussec Chronograph Open Home Time by Montblanc indicates home time using a disc rotating beneath a small offset hour and minute subdial adjustable to local time.

Hysek, *Verdict Automatic Triple Time Zone*

On the Verdict Automatic Triple Time Zone watch by Hysek, the second and third time zones are indicated by retrograde displays in the dial center, coordinated by two apertures displaying the names of cities respectively situated to the east and west of the Greenwich meridian.

Corum

Admiral's Cup Challenge 48 Day & Night

The Admiral's Cup Challenge 48 Day & Night from Corum is adorned with a map of the world that bears a transparent rotating disc. This system can perform two functions: it can display a second time using a small white indicator, or it can indicate which side of the Earth is illuminated by the sun and which is in darkness.

ochs und junior, *Due Ore*

The epitome of ingenuity and restrained aesthetics, the Due Ore watch by ochs und junior, developed by Ludwig Oechslin, displays a second time zone through apertures by means of a crown-adjustable rotating disc.

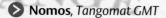

> **Nomos**, *Tangomat GMT*

On the Tangomat GMT from the German brand Nomos, the central time and the name of the reference city (indicated by abbreviated airport codes through an aperture at 9 o'clock) may be simultaneously adjusted by pressing a pushpiece at 2 o'clock, while a 24-hour second time-zone display appears through an aperture at 3 o'clock.

> **A. Lange & Söhne**, *Lange 1 Time Zone*

A. Lange & Söhne's Lange 1 Time Zone possesses two auxiliary hour/minute dials with day/night indicators. The wearer adjusts the second time zone with a pushbutton that turns the rotating city ring.

> **Vogard**, *Datezoner*

The Datezoner from Vogard stands out for its patented system that allows the wearer to change the time zone and the date simultaneously, via the bezel. One needs only to pull the security lever and rotate the bezel to select the desired city.

World Times

Queen of travel watches, the world-time complication allows the wearer to see what time it is in all 24 time zones. To this end, they are usually equipped with a mobile ring that bears the names of 24 reference cities. A few watchmakers have chosen to reinterpret this classic model by varying the configuration of the dials, taking partial time zones into account or displaying specific time slots, such as the opening of the world's stock markets. The difference also finds scope for expression in the adjusting systems, whose practicality varies with the model.

Baume & Mercier, *Capeland Worldtimer*

Baume & Mercier has enriched its Capeland collection with a Worldtimer model, a steel or red-gold watch featuring a world-time function powered by a self-winding manufacture movement visible through a transparent caseback.

Girard-Perregaux
Traveller ww.tc

The Traveller ww.tc (world-wide time control) model by Girard-Perregaux teams a chronograph with a world-time display. Its manufacture self-winding movement is housed within the pure, sleek lines of a steel or steel and titanium case.

Vacheron Constantin, *Patrimony Traditionnelle World Time*

The Patrimony Traditionnelle World Time by Vacheron Constantin simultaneously displays no fewer than 37 time zones—including those that differ from their neighbors by a half-hour or a quarter-hour. This exclusive piece is patented.

Richard Mille, *RM 58-01 World Timer Jean Todt*

On the Richard Mille RM 58-01 World Timer Jean Todt, a 35-piece limited edition dedicated to the President of the Fédération internationale de l'automobile (FIA), the main time-zone change is done by simply rotating the bezel counter-clockwise.

⌃ Antoine Preziuso, *Transworld*

Transworld, from Antoine Preziuso, offers a very attractive reinterpretation of world time, with a mobile world map in the center of the dial, depicting the Earth as seen from the South Pole and making a complete rotation over 24 hours. The display is as original as it is poetic.

⌃ Andersen Genève, *Communication 750**

On the Communication 750* by Andersen Genève, the central blue-gold dial completes one revolution per day while indicating the 24 time zones opposite the names of 30 major cities on five continents.

⌃ Patek Philippe *World Time Ref. 5131*

Heir to a long tradition, the World Time from Patek Philippe is distinguished by its dial in cloisonné enamel. To change the main time zone, the wearer uses the sole pushbutton, at 10 o'clock—the mechanism then simultaneously adjusts the city disc, the 24-hour disc and the hour hand.

< Montblanc, *TimeWalker World-Time Hemispheres*

The TimeWalker World-Time Hemispheres watch by Montblanc is available in two versions: one for each hemisphere. Its dial displays a map of the Earth seen from the North or South Poles, with a corresponding choice of cities.

< Tissot, *Heritage Navigator*

In 2013, Tissot celebrated its 160th anniversary by re-issuing a 1953 model. Equipped with a chronometer-certified movement, this Heritage Navigator watch displays the time in all 24 time zones by means of the city names appearing on the dial.

> Breitling, *Transocean Chronograph Unitime*

On the Transocean Chronograph Unitime from Breitling, driven by an in-house made movement, the wearer need only turn the crown to the front or back to correct all indications with one movement—with an automatic change in the local date.

A. LANGE & SÖHNE

LANGE 1 TIME ZONE LUMINOUS – REF. 116.039

Operating on the solid silver dial of the 18K white-gold Lange 1 Time Zone, the luminous rhodiumed-gold hands and appliqués on this timepiece guide their owner even in the darkness of night. Conveniently adjusted via a pushbutton, the peripherally rotating cities ring offers a cosmopolitan and clear view of the applicable time zone's chosen reference location. Furthermore, this manual-winding timepiece is armed with a 72-hour power reserve, and boasts an ingenious synchronization mechanism that permits the seamless swapping of the indicated time between the two subdials, both of which possess an independent day/night indicator.

A. LANGE & SÖHNE

SAXONIA DUAL TIME – REF. 385.026

Powered by a completely redesigned self-winding caliber L086.2 with a 72-hour power reserve, the Saxonia Dual Time clearly indicates two time zones on the dial. The solid gold hour hand that indicates local time can be moved forward or backward in one-hour increments by two pushbuttons located at 8:00 and 10:00. A supplementary hour hand in blued steel permanently keeps time at the place of origin. The pink- or white-gold 40mm case is fitted with an antireflective sapphire crystal on the front and back. The timepiece is presented on a hand-stitched crocodile strap with a solid gold buckle.

AUDEMARS PIGUET

ROYAL OAK CONCEPT GMT TOURBILLON – REF. 265600IO.OO.D002CA.01

Clad in state-of-the-art grade-5 titanium, this timepiece leans on impressive innovation to adorn the wrist with elegant simplicity. Driven by the remarkably finished manual-winding 2913 caliber, the watch features an original function indicator at 6:00 ("H" for hours and minutes, "R" for winding, and "N" for neutral), controlled by a three-position crown. Opposing the tourbillon at 9:00, the ingenious secondary time zone at 3:00 is comprised of two overlapping discs. While the numbered disc, which indicates 1 through 12, makes two revolutions per day, the disc beneath it, visible via a transparency effect, indicates, as a white or black backdrop, the occurrence of day or night.

AUDEMARS PIGUET

ROYAL OAK CONCEPT GMT TOURBILLON – REF. 26580IO.OO.D010CA.SDT

Bright, white, sleek and powerful, the Royal Oak Concept GMT Tourbillon is a new model, sculpted with a titanium case middle and integrated rubber strap. The white ceramic bezel frames the highly architectural array of tourbillon carriage and second GMT time-zone display within. The white center stage, an intricately machined upper bridge made of white ceramic, matches the ceramic in the movement itself, visually transforming the new caliber 2930. The tourbillon cage comprises 85 components but is only 0.45g in weight. Each component is beveled, polished, assembled and polished by hand. The GMT display provides an instant reading of the time in a second time zone. It is adjusted using the pushbutton at 4:00.

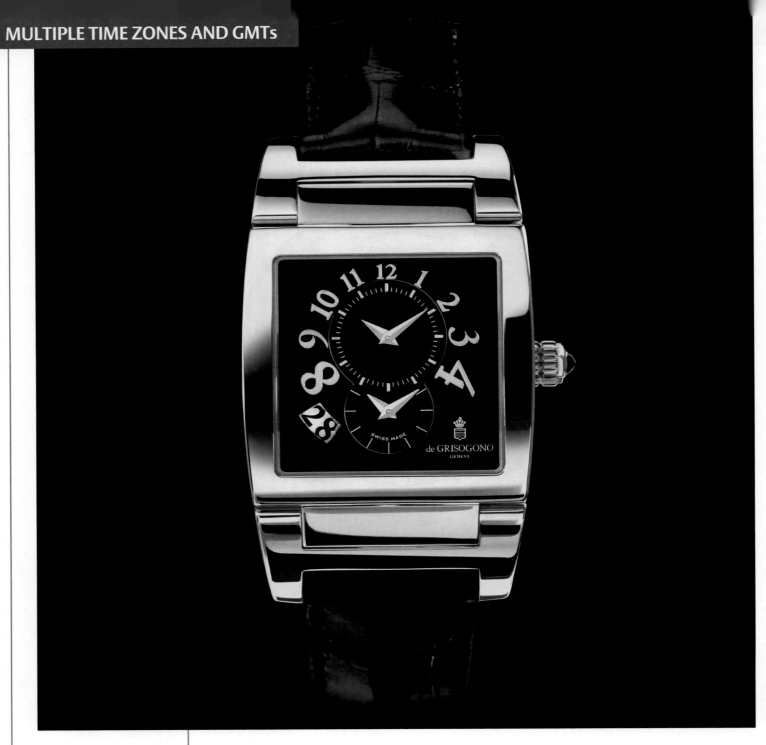

de GRISOGONO

INSTRUMENTO NO UNO – REF. UNO DF N11

This timepiece's black lacquered dial is enriched by the warmth of its case in 18K pink gold and echoed by a crown adorned with a black-diamond cabochon. On the dial, gradually enlarged 18K pink-gold Arabic numerals lead the eye from the primary time display on the upper half to the self-winding watch's secondary time zone directly beneath it. A white date aperture below the numeral "8" completes a creative and balanced geometry, and the movement may be admired through a sapphire crystal caseback.

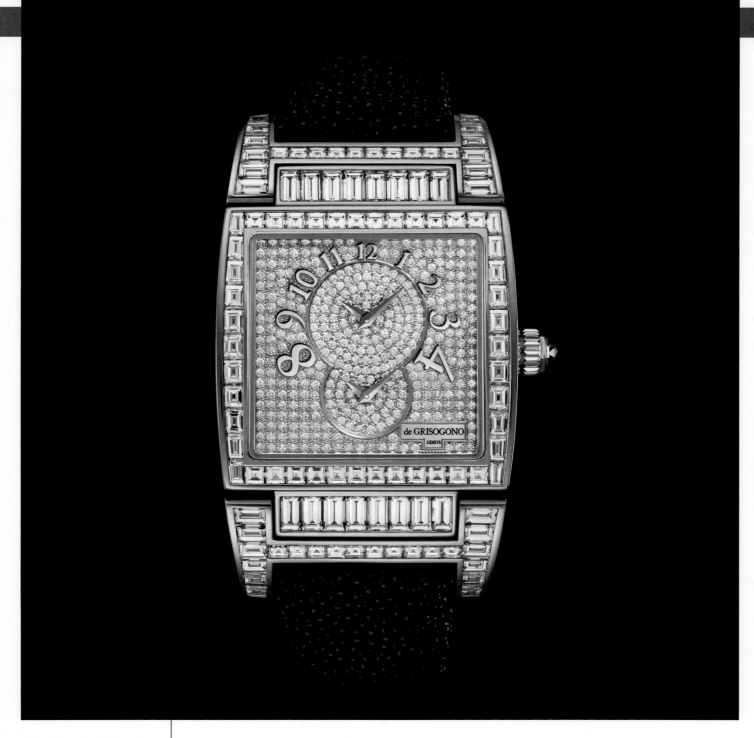

de GRISOGONO

INSTRUMENTO NO UNO – REF. UNO DF S39

Joining the shimmering splendor of the 478 diamonds set on the bezel, horns and lugs of this timepiece's 18K white-gold case, a black diamond on the crown plays off the textured color of the watch's black galuchat strap. On the dial, two separate time zones are displayed on a sea of 365 sparkling diamonds. A sequence of gradually enlarged 18K white-gold numerals adds balance to this luxurious work of horological art.

de GRISOGONO

INSTRUMENTO NO UNO – REF. UNO DF XL N02

A superb blend of rich materials and intersecting geometry, this self-winding timepiece encloses its dial within a case in satin-polished white-gold and matte black PVD-coated titanium. Framed by an outer dial in dark Macassar ebony wood, the primary silver counter, decorated with Clou de Paris, displays the hours and minutes with two dauphine hands against 12 polished white-gold indexes. A secondary time zone, in dark satin-finished silver, intersects the primary display without disrupting the legibility of either exhibition. This perfect balance is finished with a generous date window between 7:00 and 8:00.

de GRISOGONO

MECCANICO – REF. MECCANICO DG N04

This ultra-modern timepiece in titanium and 18K pink gold is driven by a 651-component hand-wound caliber that uses sophistication in the service of a display of an intuitive and legible secondary time zone. Beneath the traditional circular indication of the time at 12:00, through which the wearer can appreciate the details of the movement, a digital display is animated by the instantaneous rotation of up to 12 individual micro-segments for each transition. This avant-garde module is composed of 23 cams, a set of gears and a trigger/synchronization device.

F.P. JOURNE

OCTA UTC

F.P. Journe presents the Octa UTC, a patented system that mechanically indicates the different time periods linked with the earth's geographical positions, including summer and winter hours. On the dial, blue hands indicate the local hours, which are linked to the calendar indicating the current geographical time. The rose-gold hand indicates the hour of the second time zone on a 24-hour scale, represented by the earth's 24 time zones. The automatic-winding movement is constructed on the 1300.3 caliber, and offers an additional complication to the Octa line. The dial features off-centered hours, minutes and seconds, a retrograde power reserve, a large date, and for the first time, a dial with the earth divided in time zones. The passing of the months from 28 to 31 days is done manually.

FREDERIQUE CONSTANT

WORLDTIMER MANUFACTURE – FC-718MC4H4

Complete with a Perlage and circular Côtes de Genève decoration on the movement, this model's gleaming silver dial features an intricate guilloché design at the center, surrounded by striking, hand-polished black oxidized hands and Roman numeral hour markers. In addition to the extra-large date counter at 6:00, a 24-hour disc with day-night indicators surrounds the Roman numerals. Surrounding the outer ring of the watch face, the Worldtimer feature shows the time in 24 of the world's greatest cities, simultaneously. This model is complete with a glossy black alligator leather strap with a rose-gold pin tongue buckle.

GUY ELLIA

JUMBO HEURE UNIVERSELLE

One of the latest creations by designer Guy Ellia is the Jumbo Heure Universelle with a movement created by Frédéric Piguet. Caliber PGE 1150 features a 72-hour power reserve, a blue sapphire disc, and Côtes de Genève-finished bridges with rhodium plating. The day/night indicator, the large date and 24 time-zone indications define its singular functions. The Jumbo Heure Universelle bears an exceptional 50mm size that is rarely seen in the world of luxury watches. This model is mounted on an alligator strap with a folding buckle.

HUBLOT

KING POWER KING CASH – REF. 771.QX.1179.RX.CSH13

A sophisticated timekeeping expression of modern business travel, the aptly named King Power King Cash takes a brilliantly original approach to the GMT concept. Telling the time in multiple time zones, via four rotating aluminum discs read in conjunction with the engraved names of the world's major stock exchange markets, this 48mm carbon-fiber timepiece presents a wealth of indications thanks to its ingenious mechanical construction and unique architecture. Powered by the automatic HUB 1220 Unico movement with GMT function, visible through the skeletonized dial, this King Power wristwatch is cleverly finished with an array of green accents, including green and white SuperLumiNova in the hour and minute hands to enhance nighttime legibility.

HUBLOT

KING POWER UNICO GMT – REF. 771.CI.1170.RX

The first evolution to the exceptional Unico manufacture movement, the HUB1220 Unico base with GMT function is at the heart of this expressive timepiece with an innovative time zone complication. Controlled by a pusher at 2:00, four rotating aluminum discs simultaneously position themselves to display the time in one of 14 available cities. Within the 48mm micro-blasted black ceramic case, the black skeleton dial impressively promotes legibility, while the distinctive personality of the architecture, along with the watch's stunning mechanics, are made visible to the owner.

IWC

INGENIEUR DUAL TIME TITANIUM – REF. IW326403

This 45mm titanium wristwatch brings optimal clarity to the concept of secondary time zones. On its titanium-colored dial, joining two luminescent hour and minute hands and a white-tipped central seconds stem, an arrow-tipped indicator displays the time in a chosen location, against a peripheral 24-hour ring with contrasting shades for an unambiguous differentiation of the day/night hours. A date aperture at 3:00 completes this self-winding timepiece, which is worn on a black rubber strap with titanium pin buckle.

JACOB & CO.

EPIC SF24 – REF. 500.100.40.NS.NY.1NS

This 45mm 18K rose-gold timepiece vertically indicates a secondary time zone with an unprecedented dual split-flap system that evokes the dynamic boards of the world's most famous travel terminals. Displaying the chosen city and the appropriate hour at the top of an openworked dial with micro-blasted gray center, upon which are indicated the hours, minutes and seconds, the instantaneous 24-hour world-time mechanism provides a fascinating vintage contrast to the otherwise ultra-modern design codes of the timepiece. Boasting a 40-hour power reserve, the watch's exclusive JCAA02 self-winding caliber is finished with Côtes de Genève and circular graining.

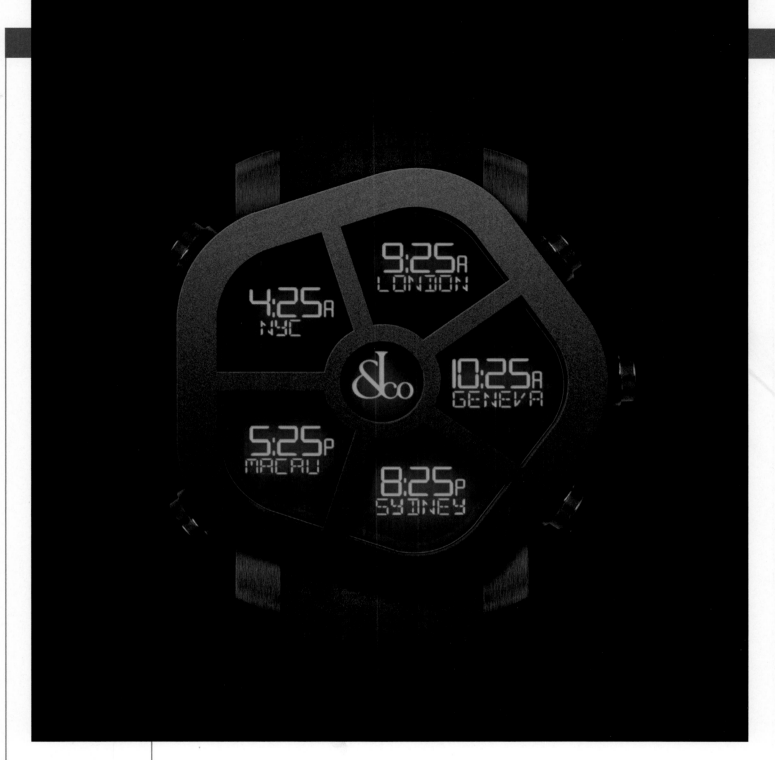

JACOB & CO.

GHOST

The Ghost timepiece from Jacob & Co. presents a new take on the GMT watch, with five mini-screens that present day, date, multiple world cities and battery reserve power. Ghost is powered by a lithium-polymer battery movement that beats at 32,768 Hz. A "mode" button allows the user to scroll through the watch's different screens, and a "color/GPS" button changes the color of the display with one short push (red, orange, yellow, green, blue, purple or white) or activates GPS localization, with more sustained pressure. The Ghost's battery can be charged via a USB cable hooked up to a computer or electrical outlet.

PANERAI

LUMINOR 1950 10 DAYS GMT CERAMICA – 44MM – REF. PAM00335

The black and opaque ceramic, long power reserve and second time zone are the distinguishing features of the Luminor 1950 10 Days GMT Ceramica – 44mm. The case on this new model is made of zirconium oxide ceramic, a very durable and scratch-resistant material. It is water resistant to 10atm and is faithful to the aesthetic of the historic model, including the crown-protecting lever bridge, which is also in ceramic. Produced in a limited edition of 500 pieces, the new Luminor model is complete with a black leather strap and fastened with a large titanium buckle customized with the Panerai logo.

PANERAI

RADIOMIR 8 DAYS GMT ORO ROSSO – 45MM – REF. PAM00395

Powered by a manual-winding Panerai P.2002/10 caliber, this model blends elegance and sophistication with a power reserve of eight days. Beating at 28,800 vph, this Radiomir piece functions with a second time zone, a 24-hour indicator and a linear power reserve indicator. The red gold used by Panerai is 5NPt, which is a special alloy with a rich color. This is used on both the cushion-shaped case as well as the wheelwork, which is revealed through the transparent caseback. This Radiomir model is a special edition piece.

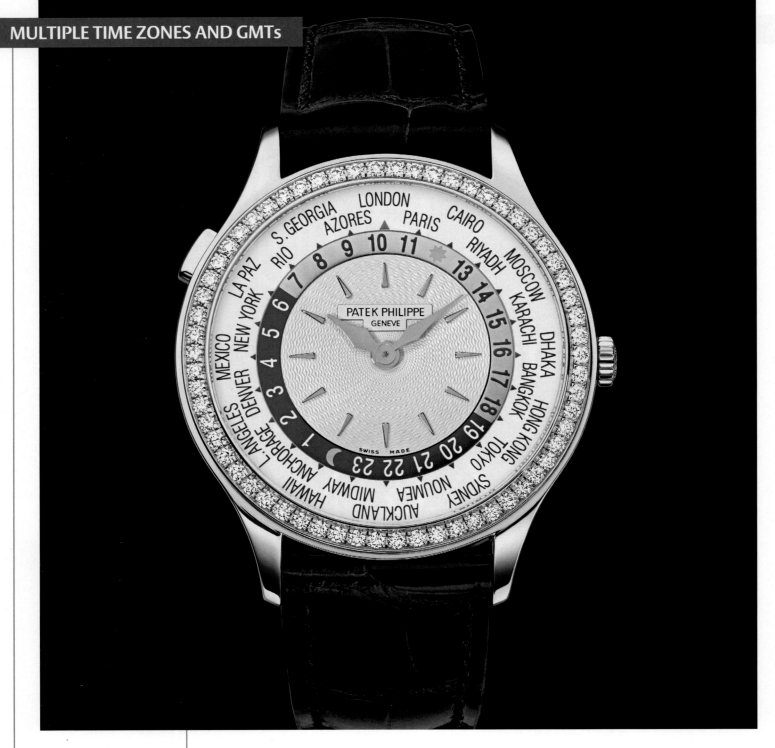

PATEK PHILIPPE

LADIES COMPLICATIONS – REF. 7130R-001

Upon the aesthetic finesse of a hand-guilloché ivory-opaline dial, this self-winding wristwatch permits its feminine wearer to travel the globe at a single glance. Displaying 24 time zones against a 24-hour disc enhanced with two-tone day/night indicator, this 36mm timepiece sparkles with 62 diamonds set upon its luxurious rose-gold case. The heartbeat of the creation, the 240 HU caliber, boasts a 22K gold off-centered mini-rotor and 48-hour power reserve. A dazzling finishing touch, the hand-stitched dark chestnut alligator strap's prong buckle is set with an additional 27 diamonds.

PATEK PHILIPPE

LADIES COMPLICATIONS – REF. 7134G-001

This white-gold timepiece set with 112 diamonds ensures that its wearer is never far from home. Driven by the hand-wound 215 PS FUS 24H caliber, the 35mm wristwatch boasts three central hands on its brown sunburst dial: two for the local hours and minutes, distinguished by their luminescent coating, and one that indicates the hour in a chosen location. Additionally, echoing the geometry of the seconds subdial at 6:00, a 24-hour "home time" indicator at 12:00 provides absolute clarity as to the time in the secondary location. The watch is worn on a hand-stitched taupe alligator strap with square scales.

PATEK PHILIPPE

MEN COMPLICATIONS – REF. 5130J-001

Housed in a 39.5mm yellow-gold case, this 24-time-zone timepiece plays off the sun-like brilliance of its frame with a sunburst silvery white dial adorned with an intricate guilloché decoration. A two-tone 24-hour disc provides instantaneous confirmation of the day/night status of the planet's numerous time zones. Animated by the 239-component self-winding 240 HU caliber with 22K gold off-centered mini-rotor, this world-time wristwatch is worn on a hand-stitched chocolate brown alligator strap with square scales.

STÜHRLING ORIGINAL

DT BRIDGE

Housed in a 42mm stainless steel case with polished coin-edge bezel, the self-winding DT Bridge accompanies its easily read secondary time zone at 9:00 with an animated AM/PM indicator at 12:00, both standing out with superb clarity above the intricate details of the dial's openworked architecture. The hours and minutes, presented with ideal contrast via two blue spade-style hands, orbit above an unobstructed demonstration of the ST-91015 caliber charismatic construction. Worn on a brown alligator leather strap, the DT Bridge boasts an exhibition caseback and a power reserve of 36 hours.

STÜHRLING ORIGINAL

EMPEROR LUSTER

Within the regal personality of its rose-tone stainless steel case with 12:00 crown, the Emperor's self-winding ST-91011 caliber is showcased in all its delicacy and sophistication through a skeletonized floating dial. An animated AM/PM display at 9:00 complements the 6:00 secondary time zone with an illustration of the sun and moon on a starry black sky. Two bright blue Breguet-style hands indicate the hours and minutes from the center of a decorated rose-tone bridge that echoes the warm tones of the 46mm case with polished coin-edge bezel. Boasting a 60-hour power reserve, the Emperor Luster's stainless steel movement may be admired further through the timepiece's Krysterna-crystal exhibition caseback.

VACHERON CONSTANTIN

OVERSEAS DUAL TIME – REF. 47450/B01R-9404

Dressed in a case and bracelet made of radiant 18K pink gold, the 1222 SC automatic movement beats in style at a rate of 28,800 vph. A pink-gold stripe at 3:00 ruptures the dial's fine finishing to indicate the remainder of the watch's 40-hour power reserve. Enhanced with luminous materials on the hours and minutes, the three-hand time display is joined by a date subdial at 2:00, and legible secondary time zone with black and white day/night indicator. On the watch's solid caseback, a stamped medallion of the Overseas sailboat tells the story of the collection's heritage.

VACHERON CONSTANTIN

PATRIMONY TRADITIONNELLE WORLD TIME – REF. 86060/000R-9640

Combining Vacheron Constantin's rich multiple time zones heritage with a driven spirit of innovation, the self-winding 2460 WT caliber takes into account the modern reality of the world's 37 time zones. By aligning the reference city with a black triangle at 6:00, the owner has immediate access to time in all of the globe's regions, including those with partial time zones, be they half or even quarter hours. Brilliantly accomplished within a 42.5mm 18K pink-gold case, this masterpiece ingeniously uses three dials to obtain such a supreme degree of complexity with optimal legibility. In addition to driving the central hours, minutes, and seconds, with a frequency of 28,800 vph, the movement precisely rotates a sapphire dial with 24-hour chapter ring and map-overlaying day/night indicator. Simplifying the use of the watch, all indication adjustments are made via a single crown. As if the front of the watch weren't spectacular enough, the Patrimony Traditionnelle World Time features a sapphire crystal openworked caseback.

ZENITH

PILOT MONTRE D'AÉRONEF TYPE 20 GMT – REF. 03.2430.693/21.C723

This self-winding tribute to Zenith's aviation history uses the generous size of its 48mm stainless steel case to display the hours, minutes, small seconds, and peripheral secondary time zone with absolute clarity and nighttime legibility. On the matte black dial enhanced with Arabic numerals made entirely of SuperLumiNova, the Elite 693 caliber, with 50-hour power reserve, animates a red-tipped hand that indicates the time in a chosen location via a peripheral 24-hour scale. On the caseback, the Zenith Flying Instruments logo pays tribute to the brand's proud aviation history.

Calendars

Patek Philippe
Celestial with Date

Calendar Watches

The calendar is undoubtedly one of the most useful functions a watch can offer—even though the widespread use of electronic diaries, computers and cellphones makes it less indispensable than in the past. It is also one of the complications that appears in the most varied guises. Today one finds a vast range of calendar watches, from models that feature only date windows (and requiring five corrections per year) to so-called secular calendars (valid almost indefinitely), with a middle ground of watches equipped with features such as day/date indications, annual calendars (just one adjustment per year) and the famous perpetual calendars (able to keep track of month and leap-year cycles). Imagination also runs riot on the design side, with models featuring extremely original layouts, innovative display modes, open-heart mechanisms and the rise of thoroughly feminine models.

STRIKING SYSTEMS WERE BORN AT THE SAME TIME AS MECHANICAL HOROLOGY.

The very first medieval monumental clocks (particularly the 14th- and 15th-century cathedral clocks) already displayed a number of astronomical indications, some of which were related to the calendar. Calendar displays appeared at a relatively early stage on pocket watches—well before the minutes and seconds hands. From the late 16th century onwards, certain astronomical watches showed the time, the date, the day the week, and the month (with its length). Calendar watches subsequently proved a great success, but they were nonetheless merely "simple calendars" requiring manual adjustment after each month with less than 31 days, meaning five times a year. It was not until the late 18th century that the so-called "perpetual calendars" were developed, taking account of the variable lengths of the months and of the leap-year cycle.

The 1920s saw the emergence of the first perpetual calendar wristwatches representing authentic feats of miniaturization. In 1945, Rolex claimed the launch of the first wristwatch with date window, the Datejust, which a few years later was fitted with a magnifying glass to facilitate reading of the date. The date window subsequently became a classic feature of "simple" watches. In the 1980s, with the rebirth of mechanical watchmaking, the perpetual calendar made a spectacular comeback as a symbol of technical mastery and expertise, and all the major brands were keen to perfect this already highly sophisticated mechanism. The start of the new millennium has witnessed the breakthrough of a complication that was curiously rare until then, the annual calendar, which is more practical than the simple calendar, but less complex and less expensive than the famous perpetual calendars.

HOW DOES IT WORK?

The Gregorian calendar poses several challenges to watchmakers, particularly the varying lengths of the months and the periodical return of February 29. There are several different kinds of mechanisms, with distinct degrees of complexity

SIMPLE CALENDAR

This is the name given to a calendar systematically indicating 31 days per month. A simple calendar must be manually adjusted on the first day of months following a month with fewer than 31 days, meaning five times a year (March 1st, May 1st, July 1st, October 1st and December 1st).

The simple calendar mechanism operates by means of a reducing gear train activated via the hour wheel. It moves one notch forward with every two turns of the hour wheel (24 hours). The display appears either on a disc bearing the date, or via a hand fixed to the 31-day star.

COMPLETE CALENDAR

A complete calendar looks very much like a perpetual calendar, in that it displays the date, the day, the month and generally the moonphase. But it does not take account of the varying lengths of the months and must therefore be corrected five times a year.

ANNUAL CALENDAR

The annual calendar automatically recognizes all months of 30 and 31 days, but does not take account of the duration of the month of February. It must therefore be corrected once a year, on March 1st. Annual calendars generally operate with the help of a wheel and pinion system of the same basic type as that driving the simple calendar, but considerably more complex.

PERPETUAL CALENDAR

The perpetual calendar is a highly sophisticated mechanism that not only recognizes months of 30 and 31 days, but also the 28 days of February and the quadrennial return of February 29th. It therefore possesses a mechanical "memory" of 1,461 days and theoretically requires no manual intervention. It does not however take account of the Gregorian correction relating to secular or century years and must be corrected for every secular year of which the first two figures are not divisible by 4—as will for instance be the case in 2100.

The perpetual calendar is distinguished from other types of calendar by the presence of a highly sophisticated toggle (or multiple-lever) system that pivots on its axis to drive the various displays. The indications may be displayed by discs or hands.

SECULAR OR CENTURY CALENDAR

The secular or century calendar takes account of the Gregorian correction made to years ending with 00. This extremely complex and rare calendar appears only on a handful of ultra-complicated watches (see the chapter on Multi-Complications).

Simple Calendars

From watches with a date window to those featuring calendars complete with day, date and month indications, there is a wide spectrum of simple calendar watches, which require correction at the end of each month lasting fewer than 31 days (i.e. five times a year). From a classic style to contemporary reinterpretation, watchmakers are giving free rein to their creativity, with an appealing selection of innovative designs, handed displays and retrograde indications—enough to satisfy any desire.

Jaquet Droz, *Grande Seconde Quantième*

On the Grande Seconde Quantième from Jaquet Droz, the date display is effected with a hand on the oversized seconds subdial at 6 o'clock, while hours and minutes share the off-centered subdial at 12 o'clock. The blue or brown dial is adorned with a Côtes de Genève motif.

Piaget, *Altiplano Date*

Equipped with an ultra-thin mechanical self-winding movement (3mm thin), the Altiplano Date by Piaget features an extremely uncluttered face enlivened by a date window at 9 o'clock and a small offset seconds subdial at 4:30.

> **Parmigiani**, *Ovale Pantograph*

Immediately recognizable by its highly original telescopic hour and minute hands, the Oval Pantograph by Parmigiani displays the date through an opening shaped like the arc of a circle at 6 o'clock. It has an 8-day (192-hour) power reserve.

> **Girard-Perregaux**, *1966 Complete Calendar*

A paragon of readability, the 1966 Complete Calendar by Girard-Perregaux indicates the day and month through two in-line apertures, while a small hand points to the date on a scale surrounding the moonphase display.

> **Meistersinger**, *Perigraph*

Equipped with a single hand for the hours and minutes, the Perigraph by the German brand Meistersinger indicates the date at 12 o'clock by a red marker pointing to the numbers on the rotating, open calendar ring.

Louis Erard, *Emotion*

On the Emotion model by Louis Erard, the date plays the starring role at 12 o'clock thanks to a disc that appears through an aperture in the guilloché dial.

Rolex, *Oyster Perpetual Day-Date II*

A worthy heir to the first wristwatch to spell out the entire name of the day (launched in 1956), the Rolex Oyster Perpetual Day-Date II offers a choice of 26 languages for this indication. The date aperture is topped by the famous "Cyclops" lens, which magnifies the numeral 2.5x.

De Bethune
DB27 Titan Hawk

The DB27 Titan Hawk from De Bethune stands out for its finely worked grade-5 titanium case, its mobile floating lugs that embrace the shape of the wearer's wrist and its date display, which uses a central hand open-worked in a very original shape.

▶ Movado
Red Label Museum Calendomatic

Inspired by a model launched in 1946, the Red Label Museum Calendomatic by Movado combines the famous understated black Museum dial with an original date display by means of a rotating disc bearing red Roman or Arabic numerals.

▼ Cuervo y Sobrinos
Robusto Day-Date Churchill

The Swiss-Cuban brand Cuervo y Sobrinos has dedicated this limited edition, with aperture date and day displays, to one of the most famous cigar-smokers of all time.

▶ Blancpain, *Retrograde Calendar*

The Retrograde Calendar from Blancpain's women's collection displays the date upon its mother-of-pearl dial using a sinuous hand tipped with a small star. It is also equipped with a moonphase and a flower-shaped oscillating weight.

Large Dates

Highly prized for their readability as well as their original touch, "large date" displays are becoming increasingly sought-after in recent years, whether appearing alone on the dial or associated with other complications. Several well-known brands have taken an interest in this ingenious mechanism, improving both its running and its reliability.

H. Moser & Cie, *Monard Date*

Based on two superimposed discs, the large date display on the Monard Date by H. Moser & Cie is also distinguished by its patented Double Pull Crown system, which avoids the user accidentally changing the time when simply seeking to adjust the date.

∧ A. Favre & Fils
Phoenix 10.1 Quantième à Grand Affichage Rotatif

On the Phoenix 10.1 Quantième à Grand Affichage Rotatif by A. Favre & Fils, the units numerals remain immobile, while a red arrow (for the dates from 1 to 9) or the 1, 2 or 3 o'clock numeral moves to indicate the date.

∧ A. Lange & Söhne
Grande Lange 1 Lumen

Large date displays are closely associated with A. Lange & Söhne. A specialist in this type of display, the German brand uses its patented system on several models, including the Grande Lange 1 Lumen with a luminescent coating that enables date read-off even in the dark.

Audemars Piguet
Tourbillon Grande Date Edward Piguet

Combining two superimposed discs and an ingenious system of partial gears, the display created by Audemars Piguet is intended to be the largest on the market. Witness this ultra-sophisticated Edward Piguet Tourbillon Large Date model.

‹ Cecil Purnell, *Grande Date*

The Grande Date from Cecil Purnell uses transparent effects, revealing the two discs of the oversized date display as well as part of the movement and the tourbillon carriage.

› Glashütte Original, *Senator Panorama Date*

The large date of the Senator Panorama Date from the German firm Glashütte Original is shown on an offset twin aperture at 04:00. The particularly restrained and readable dial is adorned with blued steel pear-type hands.

Perpetual Calendars: Classic Designs

Representing the ultimate form of calendar watch, the famous perpetual calendars are characterized by their multiple displays (date, day, month, leap-year cycle), generally enriched by moonphases. In their most classic version, they generally feature three or even four subdials arranged symmetrically in relation to the dial enter. But this restrained elegance sometimes conceals technical developments that considerably enhance the user-friendliness and the security of these ultra-sophisticated mechanisms.

▲ Blancpain, *Villeret Traditional Chinese Calendar*

To appeal to Chinese watch collectors, Blancpain has released the first wristwatch equipped with a traditional Chinese (lunisolar) calendar, with a "double-hour" indication, Zodiac signs, Chinese month and date, the five elements, the ten "celestial trunks" and leap months.

◀ Blancpain

Perpetual Calendar with Under-Lug Correctors

On its Perpetual Calendar with Under-Lug Correctors, Blancpain has developed an original system serving to integrate the correctors within the four bracelet lugs, on the wrist side. A press of the finger is enough to correct the date, day, month, leap-year cycle or moonphase display.

▼ Patek Philippe, *Ladies First Perpetual Calendar, Ref. 7140*

With its Ladies First Perpetual Calendar in rose gold, Patek Philippe presents its ultra-thin perpetual calendar—a classic from the manufacture—in a women's version for the first time.

▶ Audemars Piguet, *Jules Audemars Perpetual Calendar*

Equipped with an ultra-thin self-winding caliber, the Jules Audemars Perpetual Calendar model by Audemars Piguet is a perfect example of a classic and yet timeless design. The leap-year cycle appears in the center of the month subdial.

Perpetual Calendars: New Configurations

The design of perpetual calendars enables brands to play with various aesthetic arrangements in order to make their models stand out from traditional dials. The trend is towards a number of apertures placed either in a straight line or around the arc of a circle. Watchmakers also enjoy combining pointer- and window-type displays in sometimes highly original compositions, a sophisticated aesthetic approach that is associated with a number of technical accomplishments.

∧ F.P. Journe, *Octa Quantième Perpétuel*

The Octa Quantième Perpétuel by F.P. Journe features a highly legible display by means of three instant-jump date/day/month apertures. The indications are adjusted by a three-position crown, apart from the fast month correction, which has its own lever hidden between the strap lug at 1 o'clock.

∨ Chopard, L.U.C *Perpetual T*

Equipped with a chronometer-certified manufacture movement bearing the Poinçon de Genève, the L.U.C Perpetual T by Chopard combines a perpetual calendar with a tourbillon. The dial focuses firmly on readability, with a date appearing through an oversized twin aperture at 12 o'clock.

∧ A. Lange & Söhne
Lange 1 Tourbillon Perpetual Calendar

The Lange 1 Tourbillon Perpetual Calendar from A. Lange & Söhne presents its different indications in a very original way: a small, off-center hours/minutes subdial, large date in a double aperture, day with a retrograde hand at 9 o'clock and month on a disc that turns around the dial.

H. Moser & Cie., *Perpetual 1*

H. Moser & Cie. offers an unusual and extremely restrained take on the perpetual calendar. The month is displayed by a small central hand pointing to one of the 12 hour markers—such as 5 standing for May. The calendar may be adjusted forwards or backwards at any time.

Ulysse Nardin, *Black Toro*

The Black Toro by Ulysse Nardin stands out not only by its unusual appearance featuring a pink-gold case and ceramic bezel, but also by its correction system enabling forward or backward adjustment of all calendar indications via the crown.

Buben & Zörweg, *One Perpetual Calendar*

In addition to the day and month appearing in two in-line apertures at 12 o'clock, the One Perpetual Calendar by Buben & Zörweg provides a large date display by means of two discs that are partially visible on the dial side. It is equipped with a silicon lever and escape wheel.

Breguet, *Classique 7717BA*

On this self-winding Classique watch by Breguet, the pointer- or window-type displays are lined up along the same vertical axis, against the brand's characteristic guilloché dial.

Antoine Martin, *Quantième Perpétuel QP01*

Equipped with a silicon escapement that needs no lubrication, the Quantième Perpétuel QP01 from Antoine Martin stands out for its original aesthetic, with vertical displays for the day and month, large date and "new-style" guillochage.

Perpetual Calendars: Retrograding and Roller Systems

How can one enliven a perpetual calendar model? By choosing a retrograde-type mechanism for one or several of the calendar indications, with a hand moving across the arc of a circle before jumping back to its point of departure (see the chapter on Alternative Displays). Another, much rarer, option is that of roller-type indications. This quest for originality paves the way for all manner of fanciful visual touches calling for a superlative degree of technical mastery.

⌄ Parmigiani, *Toric Corrector*

Equipped with a minute repeater and a pusher enabling instant correction of all the calendar and moonphase functions, the Toric Corrector by Parmigiani Fleurier displays the retrograde date on the arc of a circle running from 8 to 4 o'clock.

‹ Maîtres du Temps
Chapter Two

The Chapter Two watch, created by Daniel Roth, Roger Dubuis and Peter Speake-Marin for the Maîtres du Temps brand, displays the months and days spelled out on two rollers—complete with two correctors housed on the caseback.

⌃ Piaget, *FortyFive Perpetual Calendar*

On its FortyFive Perpetual Calendar, Piaget presents a dynamic configuration, with retrograde date at 3 o'clock, retrograde day at 9 o'clock and subdial displaying the month and leap-year cycle at 12 o'clock.

⌃ Jaquet Droz, *Quantième Perpétuel*

Jaquet Droz, known for its consistently original designs, offers a Perpetual Calendar model with double retrograde display occupying the entire upper part of the dial, with days on the left and the date on the right.

⌃ Jean Dunand, *Shabaka*

The ultra-sophisticated Shabaka watch by Jean Dunand is equipped with a minute repeater striking on a cathedral chime, and is also distinguished by its roller-type display of the perpetual calendar indications, all of which jump instantly at midnight.

Perpetual Calendars: The Transparent Approach

Visible mechanisms are all the rage in the watch industry at the moment. After introducing sapphire crystal casebacks on virtually all their complicated models, brands are vying with other to find ingenious means of revealing all or part of the movement through the dial side. Perpetual calendars are no exception, and there is a vast range of models with openworked dials, skeletonized movements or transparent dials that now hide almost nothing of their inner mysteries, at least in visual terms. These spectacular stagings are often matched by some exceptional technical feats accomplished behind the scenes.

Patek Philippe
Grand Complication Ref. 5104

On the Grand Complication Ref. 5104 model by Patek Philippe, the days, months and the leap-year cycle number are printed directly on the sapphire crystal dial and dark "pallets" highlight the appropriate white transfer by contrast.

Parmigiani
Tonda 42 Retrograde Perpetual Calendar

On the Tonda 42 Retrograde Perpetual Calendar by Parmigiani, the abbreviations of the day and month appear on two discs featuring original shapes and visible through the transparent dial.

Jaeger-LeCoultre
Master Eight Days Perpetual SQ

The Master Eight Days Perpetual from Jaeger-LeCoultre, endowed with an eight-day power reserve, is shown here in a dazzling skeletonized version featuring daringly curved bridges echoing the lines of longitude and latitude.

Vacheron Constantin,
Malte Retrograde Perpetual Calendar Openface

Vacheron Constantin has equipped its Malte Retrograde Perpetual Calendar Openface watch in platinum with a double sapphire crystal. Transparent subdials with white rotating "pallets" serve to indicate the day, month and leap-year cycle painted in black on one of the sapphire crystals.

Cartier, *Tortue Perpetual Calendar*

Equipped with an in-house movement, the Tortue Perpetual Calendar by Cartier combines various types of display—including a central pointer-type date indication and a retrograde day running from 8 to 4 o'clock—appearing on an openworked dial.

Annual Calendars

Less complex than a perpetual calendar, the annual calendar enables brands to create watches that are more affordable and simpler to operate, while providing a degree of user-friendliness far superior to simple calendars in that they require just one correction per year. The last few years have brought a flourishing spate of annual calendar models featuring a great variety of displays and designs, as well as occasionally associating this complication with other useful functions.

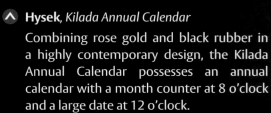

∧ **Hysek**, *Kilada Annual Calendar*

Combining rose gold and black rubber in a highly contemporary design, the Kilada Annual Calendar possesses an annual calendar with a month counter at 8 o'clock and a large date at 12 o'clock.

∧ **Parmigiani**, *Tonda Retrograde Annual Calendar*

On Parmigiani's Tonda Retrograde Annual Calendar, the date is displayed using a retrograde hand traversing an arc from 8 to 4 o'clock, while the day and month (as a number from 1-12) appear in two small apertures.

Pierre DeRoche, *GrandCliff QA Power Reserve*

The GrandCliff QA Power Reserve model by Pierre DeRoche, equipped with a Dubois Dépraz movement, combines the annual calendar with a chronograph featuring retrograding 60-minute and 6-hour counters, a flyback function and a large date.

Zenith, *Captain Winsor Annual Calendar*

Equipped with the automatic-winding El Primero caliber, Zenith's Captain Winsor Annual Calendar associates the functions of a 1/10-of-a-second chronograph with an annual calendar boasting three apertures for the date, day and month.

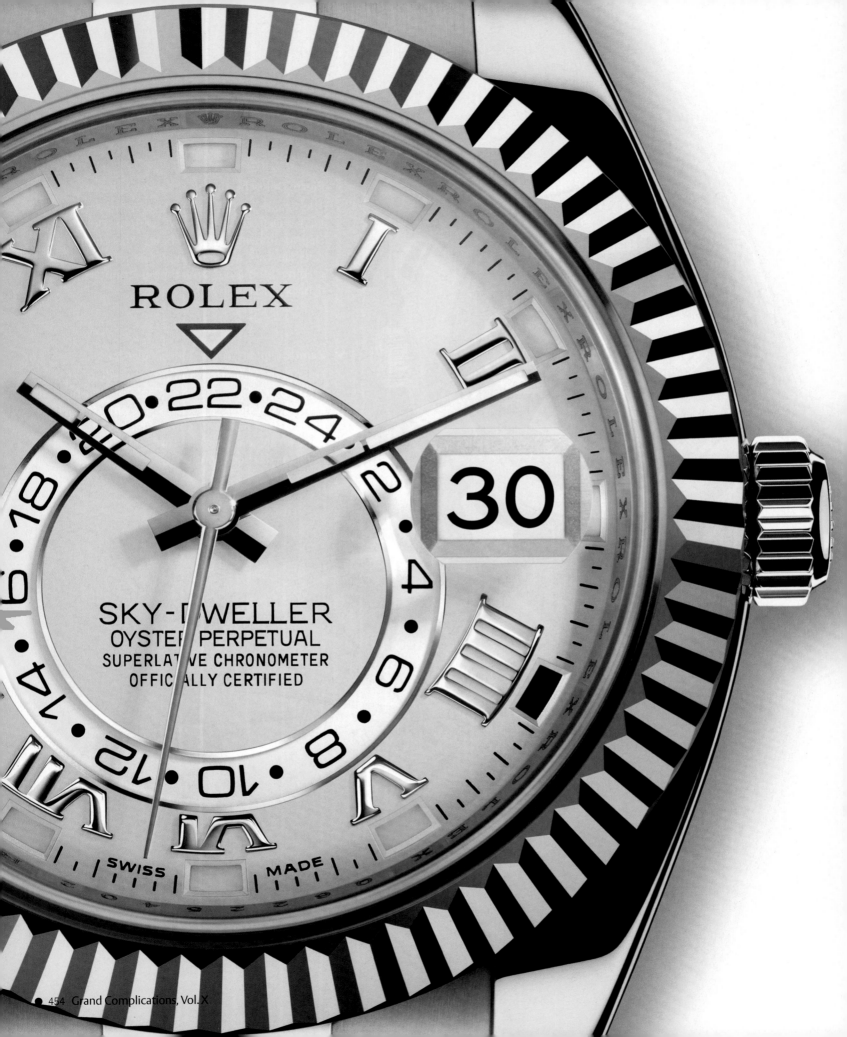

Rolex, *Oyster Perpetual Sky-Dweller*

Rolex's travel watch Oyster Perpetual Sky-Dweller pairs a dual time zone with an annual calendar. The month is displayed discreetly in 12 apertures arranged around the dial. Setting the time zones and calendar is done using the crown.

Girard-Perregaux

Cat's Eye Annual and Zodiac Calendars

Designed for women, the Cat's Eye Annual and Zodiac Calendars watch combines a snail-shaped date display with a large comet's tail-shaped aperture revealing the ever-circling dance of the Zodiac signs.

Patek Philippe

Annual Calendar, Ref. 4936

Launched in 1996, the patented Annual Calendar from Patek Philippe has become a bestseller for the brand. It has been interpreted in different versions, including the reference 4936 ladies' diamond-set model with mother-of-pearl dial.

Cartier

Rotonde de Cartier Annual Calendar

The Rotonde de Cartier Annual Calendar is equipped with a large date in a double aperture at 12 o'clock and a rapid-setting system for the calendar, governed by the crown. The days and months are indicated on two concentric rings by two hands, of which only the red tips are visible.

ochs und junior, *Anno*

The epitome of minimalism, the contemporary Anno from ochs und junior displays the calendar indications using three circular scales: date (1 to 31) around the edges of the dial, day (1 to 7) on the central axis, month (1 to 12) above the hour marker at 6 o'clock. The annual calendar mechanism designed by the famous Ludwig Oechslin has only three moving parts—as opposed to 40 in a conventional system.

Hermès, *Dressage Annual Calendar*

The Dressage Annual Calendar watch by Hermès, with its exclusive Vaucher Manufacture Fleurier movement, is distinguished by its pointer-type 270° retrograde date display and its two visible day and month discs.

A. LANGE & SÖHNE

DATOGRAPH PERPETUAL – REF. 410.032

This model is driven by the manually wound Lange L951.1 movement. It is a flyback chronograph with a precise jumping minute counter as well as perpetual calendar with a large date, moonphase display, day of the week, month, leap-year display, small seconds hand with stop seconds, and day/night indicator. The plates and bridges are made of silver and the balance cock is engraved by hand. The three-piece rose-gold case features an untreated German antireflective sapphire crystal and caseback, as well as a main pushbutton for simultaneously advancing all calendar displays, recessed pushbuttons for separately advancing the calendar displays and two chronograph pushbuttons. The hand-stitched crocodile strap is secured with a solid gold buckle.

A. LANGE & SÖHNE

GRAND LANGE 1 – REF. 117.028

With a sleek black solid silver dial and the classic Grand Lange 1 dial configuration framed by a thinner white-gold case, this piece is driven by the manufacture caliber L095.2, whose 72-hour power reserve is provided by a single mainspring barrel, reducing the thickness of the iconic piece. Rhodiumed gold appliqués on the dial, paired with luminescent hands, make this Grand Lange 1 highly legible at all times of day or night. The movement boasts several manufacture touches, such as a classic screw balance, a balance spring developed and produced in-house, a three-quarter plate in untreated German silver and lavishly hand-decorated components.

A. LANGE & SÖHNE

GRAND LANGE 1 "LUMEN" – REF. 117.035

The Lange outsize date, combined with an off-centered dial, became an emblem of the manufacture soon after its introduction. The technical secret to such an imposing indication, however, remained hidden behind the solid silver dial...until now. Parts of the dial have been replaced by semi-transparent sapphire crystal—only the outer ring and the subdials for seconds and hours and minutes are in blackened silver—revealing the disc mechanism behind the date indication. The urge to show all pervades the Grand Lange 1 "Lumen": in addition to the transparent dial, the piece also boasts luminescent indications for easy night readoff, as well as a sapphire crystal caseback that exposes the manual-winding manufacture caliber L095.2. The timepiece is available in a limited edition of 200 watches in platinum cases.

A. LANGE & SÖHNE

LANGEMATIK PERPETUAL – REF. 310.025

Powered by the self-winding Caliber L922.1 SAX-0-MAT, the Langematik Perpetual is the world's first self-winding wristwatch with a large date and a perpetual calendar featuring separate pushbuttons for individual corrections of the calendar indications (date, day of the week, month, leap year, and moonphase displays), as well as a main pushbutton for advancing all displays. Caliber L922.1 SAX-O-MAT has a 46-hour power reserve, stop seconds with a patented zero-reset mechanism, plates and bridges made of untreated German silver, and a hand-engraved balance cock. The Langematik Perpetual's three-piece platinum case is fitted with antireflective sapphire crystal on the front and back, and a crown for winding the watch and setting the time. The solid silver and argenté dial is topped with luminous platinum hands. The watch's crocodile strap is secured via a solid gold buckle.

BLANCPAIN

TRADITIONAL CHINESE CALENDAR – REF. 00888-3631-55BB

This model is based on fundamental principles established for millennia and profoundly rooted in Chinese tradition. On its dial, the hours, minutes and the Gregorian calendar rub shoulders with the main indications of the Chinese calendar: traditional double-hour indication, day, month with indication of leap months, signs of the Zodiac, the five elements and the 10 celestial stems. The combination of the latter with the 12 animals of the Zodiac follows the sixty-year cycle that is central to Chinese culture. The moonphase, a key element in Blancpain complete calendars, is also presented, and plays a particularly important role in this model, given the link between the lunar cycle and traditional Chinese months.

DE BETHUNE

DB25 QP DE BETHUNE PERPETUAL CALENDAR

The De Bethune vision of a perpetual calendar is imbued with poetry and combines horological finesse with cutting-edge mechanisms. For this model, De Bethune has developed the self-winding Calibre DB 2324 QP, which includes a perpetual calendar, a spherical moonphase display featuring a "1 day in 122 years" level of accuracy, a double self-regulating barrel, a triple pare-chute shock-absorbing system and a titanium and platinum balance. The sophistication of the décor, complete with a hand-guilloché silver-toned dial and its 12 shining sectors, accentuates the legibility of the complications. Representing an avant-garde mechanism within a poetic timepiece, the DB25 QP is crafted in the De Bethune workshops in harmony with the grand watch making tradition.

F.P. JOURNE

QUANTIEME PERPETUEL

The Quantième Perpétuel presents a distinctive aesthetic through an uncluttered dial with apertures for the day, date and month. The instantaneous jump is reinforced by an ingenious F.P. Journe system designed to accumulate energy, release it instantaneously when the date, day or month changes, then slow it down at the end of its route. The varying lengths of the months are automatically taken into account, and the leap-year cycle is displayed at the center of the dial. The automatic-winding movement, constructed on the basis of the exclusive Octa calibre 1300.3, is manufactured in 18K rose gold and possesses 120 hours of power reserve. The Quantième Perpétuel is available in platinum or in 18K red gold, with 40mm or 42mm diameter.

GLASHÜTTE ORIGINAL

SENATOR PERPETUAL CALENDAR – REF. 100-02-22-05-05

Made by hand in the Glashütte manufacture, the Senator Perpetual Calendar exemplifies the ongoing quest for the perfect union of mechanical complications and art. All Senator Perpetual Calendar displays are arranged to form a clear, harmonious balance. The day, month and Glashütte Original Panorama Date display black numerals against a white background, and the warm silver dial shows a silver moonphase display on a blue disc. A discreet, unambiguous leap-year display presents a red dot to indicate a leap year, followed by yellow, black and white dots for successive years. The heart of the Senator Perpetual Calendar, visible through the sapphire crystal caseback, is the Glashütte Original Caliber 100-02, a 28,800 vph automatic movement featuring a screw balance with 18 weighted golden screws and a power reserve of more than 55 hours.

GLASHÜTTE ORIGINAL

SENATOR PERPETUAL CALENDAR – REF. 100-02-25-12-05

On this stainless steel, black-dial version of the Senator Perpetual Calendar, the moonphase appears on a black disc adorned with silver stars and a radiant silver moon. The timepiece's reset mechanism allows easy synchronization of the second hand with a time standard. Unlike other reset mechanisms, the second hand is connected neither to the winding stem nor the crown. Consequently, when the crown is pulled out, the balance remains in oscillation and the movement continues to run, significantly reducing material stress. Pressing a separate button on the side of the case at 8:00 activates the reset mechanism. Exquisitely finished components enhance the artistry of the Glashütte Original Caliber 100-02's design: the signature Glashütte three-quarter plate with Glashütte ribbing, skeletonized rotor with 21K gold oscillation weight, twin spring barrels and swan-neck fine adjustment.

GUY ELLIA

TIME SPACE QUANTIEME PERPETUEL

The Time Space, original ambassador of Guy Ellia, is transformed into a perpetual calendar for this version. A mechanical hand-winding Frédéric Piguet caliber harbored in a 46.8mm black-gold case allows the owner to read the hour, minute, day, month, date and moonphases, while taking leap years into account. With a brushed bottom plate of 7.75mm, this watch symbolizes elegance, complexity and thinness. A specific automatic winder is integrated to the watch box. This model is also available in white or pink gold.

HUBLOT

ANTIKYTHERA SUNMOON – REF. 908.NX.1010.GR

This extremely sophisticated timepiece pays homage to humanity's first "astronomical calculator," dating back to the second century BCE. Powered by a miniaturized, simplified version of the ancient mechanical masterpiece, the 20-piece limited edition wristwatch boasts seven dial-side complications. Along with its flying tourbillon, whose one-minute rotation is used to indicate the seconds, the manually wound, 295-component movement permits its wearer to determine, for any given day, the constellations behind both the sun and the moon, displaying their sidereal positions as seen from Earth, as well as the date via the sun hand on the dial's outermost circle and the moonphase through a circular central window.

IWC

INGENIEUR PERPETUAL CALENDAR DIGITAL DATE-MONTH – REF. IW379201

Inspired by the high-tech materials of the motor sport world, IWC presents its first timepiece with a case made of titanium aluminide, an alloy lighter and tougher than pure titanium. The self-winding wristwatch echoes this innovative exploit with an inventive aesthetic approach to the perpetual calendar. Indicated at 9:00 and 3:00 by way of two large double-digit apertures, the date and month are perfectly differentiated from the flyback chronograph's intuitive hours/minutes totalizer at 12:00, and the small seconds/leap-year subdial at 6:00. The 89802 caliber is endowed with a sophisticated quick-action switch that stores a small amount of energy every night, only to release it at the end of the month and year, permitting the simultaneous activity of the calendar's numerous indicators without affecting the movement's accuracy.

PATEK PHILIPPE

GONDOLO 8 DAYS, DAY & DATE INDICATION – REF. 5200

Housed in a cambered 18K white-gold case, this hand-wound timepiece owes its impressive eight-day power reserve to the 28-20 REC 8J PS IRM C J caliber's integration of two in-line twin mainspring barrels. Patek Philippe demonstrates its architectural virtuosity on the timepiece's blue brass wafer dial embellished by a sunburst effect. While the top of the face is devoted to a highly legible power-reserve indicator, a multi-function subdial on the lower half incorporates the running seconds, date via a hand distinguished by an unmistakable red tip and instantaneous day through a generous white aperture.

PATEK PHILIPPE

MEN COMPLICATIONS – REF. 5205R-010

Within the rich contrast of its luxurious rose-gold case and profound black lacquered dial, this self-winding wristwatch presents a perfectly balanced annual calendar. While the day, date and month are easily read through three white openings between 10:00 and 2:00, the 347-component 324 S QA LU 24H caliber additionally drives a subdial at 6:00, showcasing a colorful moonphase surrounded by a hand-guided 24-hour indicator. The 40mm timepiece is finished with a sweep seconds hand and is worn on a hand-stitched alligator strap with square scales.

PIAGET

PIAGET EMPERADOR COUSSIN PERPETUAL CALENDAR – REF. G0A33019

The Piaget Emperador Coussin Perpetual Calendar is powered by the automatic-winding, ultra-thin Piaget 855P caliber. With a power reserve of 80 hours, this piece oscillates at 21,600 vph. The central hours and minutes display is accompanied by a small seconds counter at 4:00. Additionally, the perpetual calendar is complete with a retrograde date indicator at 3:00, a retrograde day indicator at 9:00, a month and leap year cycle indicator at 12:00 and a day/night indicator. This model is also equipped with a second time zone indicator at 8:00. All of these complications stand out on the blue dial that is framed by an 18K pink-gold case.

ROGER DUBUIS

LA MONEGASQUE PERPETUAL CALENDAR – REF. RDDBMG0006

This timepiece is presented in a 44mm pink-gold case with a titanium bezel coated in black DLC. With its 4Hz frequency of oscillation (28,800 vph), the 364-component RD821J caliber, with a 48-hour power reserve, is adjusted in five positions as it delivers the comprehensive indications of this perpetual calendar. Upon the rhodium-plated satin sunburst dial, La Monégasque Perpetual Calendar contains a lower half composed of a snailed date counter and seamlessly integrated moonphase. On the upper half, two gold windows set the frame for the day (written in French) and month. Completing the calendar, a leap year indicator appears discreetly at 3:00. Finished with pink-gold applied Arabic numerals, this watch features a sapphire crystal caseback.

VACHERON CONSTANTIN

PATRIMONY CONTEMPORAINE PERPETUAL CALENDAR – REF. 43175/000R-9687

On its silver-toned opaline dial with baton- and triangle-shaped 18K pink-gold hour markers, this elegant timepiece presents a perpetual calendar so effortless that the moonphase at 6:00 seems to conduct the entire time orchestra. Via a month dial at 12:00, Vacheron Constantin provides an optimally unambiguous display of the leap-year indicator residing in its interior. Contained within the 41mm 18K pink-gold case, the Poinçon de Genève-honored 112 QP self-winding movement beats at a frequency of 19,800 vph, with a 40-hour power reserve. Turned over, the openworked back can be admired through a sapphire crystal.

ZENITH

PILOT MONTRE D'AÉRONEF TYPE 20 ANNUAL CALENDAR – REF. 87.2430.4054/21.C721

Housed in a 48mm titanium case with 18K rose-gold bezel, this self-winding timepiece, powered by the 36,000 vph El Primero 4054 caliber, boasts the extraordinary combination of a chronograph and annual calendar. Needing only one adjustment per year, in the transition from February to March, the 341-component movement's calendar displays the date through a window at 6:00 and the day and month via a double aperture at 3:00. A feat of innovation, this timepiece's calendar is composed of only nine moving parts, as opposed to the 30 or more required by conventional constructions of the complication.

Moonphase

Glashütte Original
PanoMaticLunar

Pierre Jacques

CEO of De Bethune

"We prioritize quality over quantity"

In 2002, a visionary collector, David Zannetta, and a brilliant watchmaker, Denis Flageollet, decided to turn their horological aspirations into reality. Thus was born De Bethune, named for a knight who was famous for having perfected certain watch mechanisms in the 12th century. Pierre Jacques, CEO, has since joined the brand, becoming its essential third pillar. **THIS ORIGINAL TRIUMVIRATE OFFERS A NEW KIND OF HOROLOGY THAT EXISTS IN THE SPACE BETWEEN ART AND SCIENCE.**

How would you describe De Bethune?
Three concepts breathe life and inimitable style into our watches: tradition, innovation and design. Above all, our watches celebrate the rules of traditional watchmaking, including finishings. With a work force of 60 artisans, our firm completes production entirely in-house, including movements and dials, and even cases and hands. Secondly, our watches house the latest technical advances and incorporate extremely advanced materials. In ten years, our research and development department, directed by Denis Flageollet, Technical Director and Co-Founder, has led us to file nine patents, present 17 worldwide innovations and create 15 in-house movements. Finally, David Zannetta, President and Co-Founder, designs all of our watches. Their unique architecture is recognizable at a glance as a futuristic reinterpretation of traditional shapes.

‹ DB 28 Skybridge

We reaffirmed our status as a watchmaking firm. Within a philosophy of perfectly controlled growth, we increased production volume: 403 watches, up from 340 in 2012. New models have strengthened our DB 25 and DB 28 lines, while the latter, which is sporty and technical, is gradually becoming a brand icon. Along these lines, the DB 28 Skybridge melds aesthetic research, avant-garde technique and a tribute to horological tradition. Its concave dial alludes to 18th-century clocks, just as its round case evokes fob watches. It is crafted in titanium and its architecture seems to project it into the future. Suspended ball bearings mark the hours, and an arrow extends the spherical moonphase display.

The DB 28 Black Collection was also released, in a mirror-polished version as well as a matte, satin-finished version. In a notable feat, the cases and the lugs are crafted in zirconium. The watch features multiple shimmering effects from the light playing over the material and the different finishings.

In collaboration with Michael Tay (The Hour Glass), we created the DB 25 Imperial Fountain. It was inspired by the remains of the Imperial Fountain in Beijing: 12 statues of the Chinese astrological signs. We produced four sets of 12 watches each. Each engraving bears grand feu enamel.

In the realm of high complications, the DB 16 proclaimed our expertise, with a movement boasting a tourbillon, moonphase, perpetual calendar and chronometric precision.

What is the average price of your watches?
Our average price has increased over the last three years. In 2013, it was about 90,000 Swiss francs, despite some products well below that average, such as the DB 10 (25,000 Swiss francs) and the DB 27 (35,000 Swiss francs). We occupy a very high-end segment of the market, a niche within a niche!
We aim to make the most beautiful watches possible, with the best technology and materials. The present structure of our facilities could allow us to produce many more watches. However, the brand has turned a profit since 2009 and we prioritize quality over quantity. Furthermore, if we produced more than 500 watches a year, we would not be as exclusive. We want our production to remain a bit restricted. Higher volumes would take us to another level, implying certain necessities vis-à-vis marketing, public relations and sponsorships. And needless to say, the global economic situation inspires caution. We prefer to maintain a demand that is greater than the supply, and to preserve our independence!

‹ DB 25 Imperial Fountain

What new releases do you have in store for 2014?

We are building up our flagship collections, DB 25 and DB 28. The DB 25 proposes a marine tourbillon and the DB 28 in titanium possesses a digital display. In March, we launched a five-handed DB 28 tourbillon chronograph. At the same time, as is the case with the DB 16, we are entering the world of grand complications with an exceptional Dream Watch.

What are De Bethune's strengths?

The small size of our facilities allows us to react quickly and thus deliver orders in a short time. For example, we were able to deliver all of 2013's new releases before the year was up. At this level of complications and finishings, that is a relatively rare achievement. This year, we expect to provide June delivery of models that we introduced in January. Another strength is that three people lead the brand. We combine our talents at the outset of each project. David Zanetta embodies the creative visionary, Denis Flageollet is the brilliant watchmaker, while I know the market very well, and manage marketing and distribution. De Bethune is doubly independent. Firstly, we produce all the necessary components in our own facilities. Secondly, no financial group tells us what to do. We can act freely, on our own. Today, may brands claim to be financially independent, but few are true manufactures like De Bethune. For example, we can confidently guarantee long-term after-sales service for our clients.

How will you improve in the years to come?

We still don't have the kind of high profile that customers find so reassuring. However, we are on the right track: in 2011, our DB 28 Titanium won the Aiguille d'Or at the Geneva Watchmaking Grand Prix (GPHG). This highest honor showed everyone that our process leads to solid results.

Where do you find your best sales numbers?

Out of, say, 400 watches, 80 are sold in Singapore, 60 in Moscow and about 20 in the Ukraine. The balance goes to other Asian countries, Europe, the US, Mexico and South America. We've made inroads in Hong Kong and China, without particularly focusing on those regions, unlike many other brands. Thanks to this strategy, we have a slight head start in the other markets!

⌃ DB 28 Black

‹ DB 25 Imperial Fountain Dragon

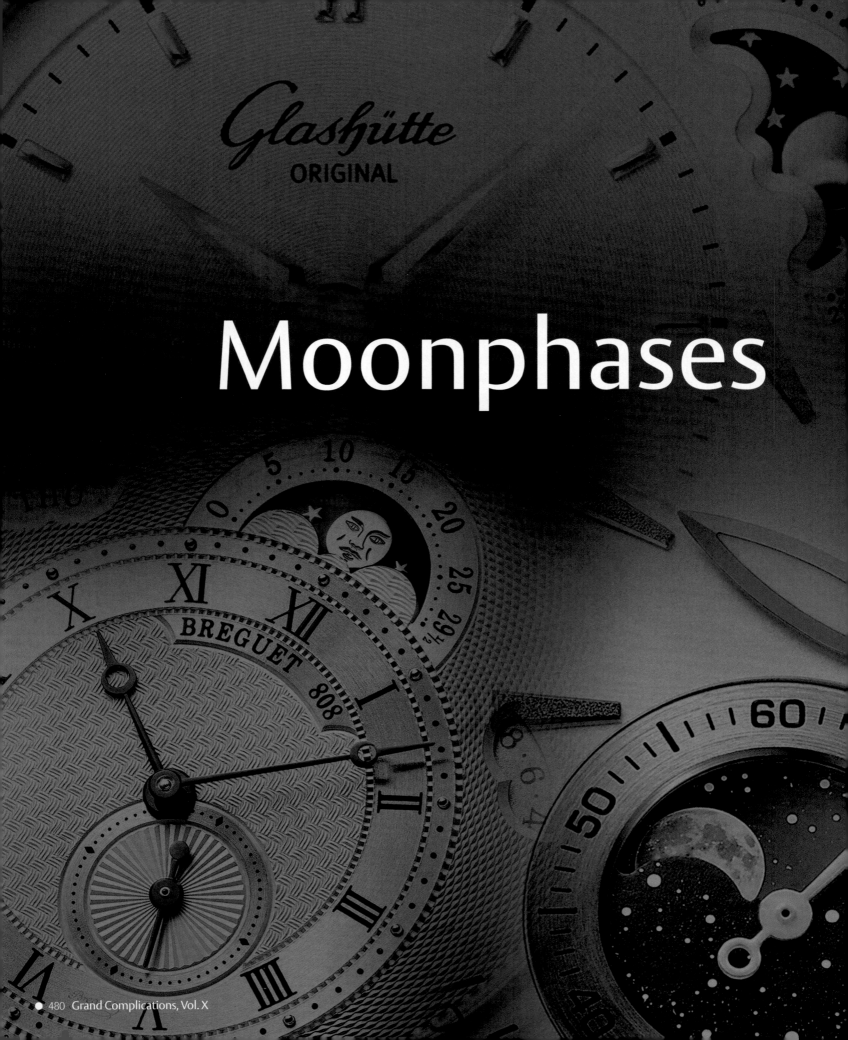

Moonphases

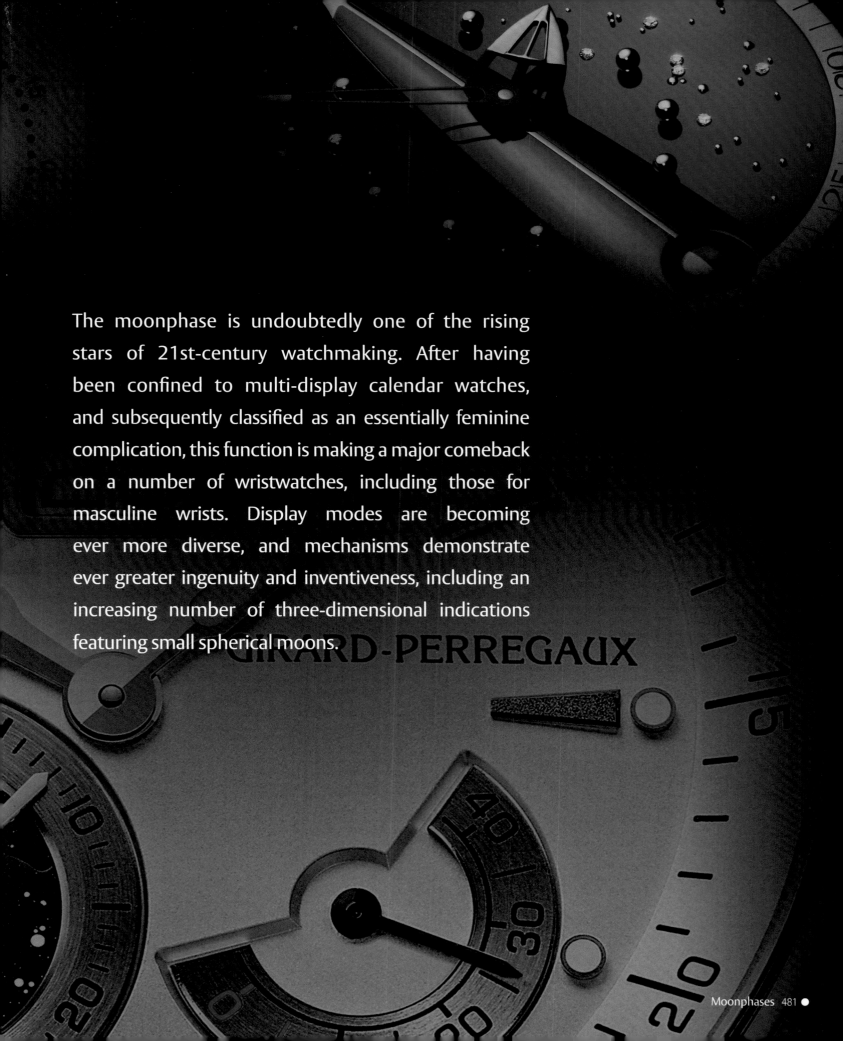

The moonphase is undoubtedly one of the rising stars of 21st-century watchmaking. After having been confined to multi-display calendar watches, and subsequently classified as an essentially feminine complication, this function is making a major comeback on a number of wristwatches, including those for masculine wrists. Display modes are becoming ever more diverse, and mechanisms demonstrate ever greater ingenuity and inventiveness, including an increasing number of three-dimensional indications featuring small spherical moons.

THE OBSERVATION OF THE MOON'S CYCLE HAS ALWAYS CONSTITUTED— ALONG WITH THE COURSE OF THE SUN, THE RETURN OF THE SEASONS AND THE ALTERNATING OF DAY AND NIGHT— ONE OF THE PRIMARY FOUNDATIONS OF CALENDAR SYSTEMS.

This cycle famously inspired the ancient Egyptians to divide the year into twelve months, a system we inherited via the Greeks and Romans. Despite the near-universal adoption of the Gregorian calendar, essentially based on the course of the sun, the ancient lunar or luni-solar (i.e. based on both the sun and moon) calendars retain their full importance in certain cultures, and are notably used to determine the date of traditional or religious holidays, or for astrological calculations. This is the case, for example, in the Chinese and Jewish luni-solar calendars. Such is also the case for the Muslim calendar, in which the year contains twelve lunar months of 29.53059 solar days—adding up to about 11 days fewer than the Gregorian calendar.

In the West, the moonphase display was an essential component of the first great clocks of the Middle Ages. It plays a role on the clock of Saint-Etienne Cathedral in Bourges, in France (1424), and on Prague's Old Town Hall (15th century). It is featured on the dials of the first pocket-watches, from the 16th century—well before watchmakers could present the minute or the second. The moonphase later became an inescapable element of watches devoted to astronomical, astrological or calendar indications—especially the famous perpetual calendars. In the 20th century, although the moon had lost much of its magnetic force over the human mind, watchmakers took it upon themselves to miniaturize the mechanism to the wristwatch format, while conserving a maximum of precision.

SWISS

HOW DOES IT WORK?

The moonphase is generally displayed via a mobile disc on which are depicted two perfectly round moons. A specially shaped window, with two rounded cutouts representing the "dark" side of the moon, allows the display to represent the diverse aspects of the heavenly body throughout its cycle (waxing, full, waning and new moons). The indicator changes once a day, as a rule.

A VARIABLE LEVEL OF PRECISION

Though they are all based on the same principle, not all window moonphase displays offer the same level of precision. Their exactitude depends to a large extent on the number of teeth on the moonphase disc. The length of one synodic, or lunar, month (the length of time from one new moon to another) is actually 29 days, 12 hours, 44 minutes and 2.8 seconds, or 29.53 days—making it impossible to simply multiply any units from solar civil time. The more teeth the disc has, the more precise the moonphase will be.

ORDINARY MOONPHASES

On ordinary moonphases, the disc bearing the two moons is linked to a wheel with 59 teeth, corresponding to two lunar months (2 x 29.5 days). This system falls behind 44 minutes and 2.8 seconds per month, or close to nine hours per year, a discrepancy that reaches a full day after two years, seven months and around 20 days.

ASTRONOMICAL MOONPHASES

High-end watches are equipped with a more complex device, called "astronomical moon," with a 135-toothed wheel. With this system, the lag between the mechanism and the actual lunar cycle comes to just one day after 122 years and 44 days. Certain recent timepieces have surpassed even this feat with impressive displays of precision (see the "Precision Records" section).

DOUBLE MOONPHASES

Some timepieces feature a simultaneous indication of the moonphase in both hemispheres. This particular feature is generally not a technical advance in and of itself (since it requires only revealing the two moons already present on most mobile discs), but it nonetheless gives the dial a charming touch of originality.

THE MOON'S AGE

Moonphases are sometimes completed by the indication of the moon's "age," i.e. the number of days that have passed since the last new moon.

Calendars with Moonphase

The moonphase indication has long been an almost inescapable element on complete or perpetual calendars (see chapter on Calendars). On these pieces, it most often occupies a small window at 6 o'clock. However, numerous brands have now chosen to reinterpret this tradition with more creative designs, moons at 12 o'clock, oversized windows or off-center displays.

< Jaeger-LeCoultre
Duomètre à Quantième Lunaire

The Duomètre à Quantième Lunaire, from Jaeger-LeCoultre, with the Dual-Wing movement, features a moonphase window complemented by two hands, one for the moon's "age," the other for the moonphase in the Southern Hemisphere.

> Breguet, *Classique 7337*

Breguet's Classique 7337 associates an off-center hour/minute subdial with two windows for the date and day. The moonphase window at 12 o'clock also displays the "age" of the moon, i.e. the number of days since the last new moon.

Dubey & Schaldenbrand
Grand Dôme DT

In designing the dial of the Grand Dôme DT, Dubey & Schaldenbrand arranged multiple displays—date, day, month, chronograph counters and moonphases—on a vertical axis, creating a highly original geometry.

Harry Winston, *Premier Perpetual Calendar*

On Harry Winston's Premier Perpetual Calendar, the moonphase shines at 6 o'clock on the off-center hours/minutes subdial, while the retrograde displays of the month and date compose a perpetually moving arrangement.

Epos, *3391 Moonphase*

On a midnight blue dial sprinkled with stars, the 3391 Moonphase from Epos displays—in an unusual configuration—a moonphase, along with windows for the day and month, and a hand for the date.

Chopard, *L.U.C Lunar One*

The perpetual calendar L.U.C Lunar One from Chopard offers an "orbital" moonphase: the small round window, in which the silver moon waxes and wanes, itself rotates around the small seconds axis on a disc that simulates the star-spangled night sky.

The Moon is the Star

Over the last several years, the moonphase indication has been breaking free of the perpetual or complete calendar "straitjacket," and establishing itself as a complication in its own right. On certain timepieces, it is even the main attraction—with displays that can be quite poetic.

> **Glashütte Original**, *PanoLunarTourbillon*

The original dial layout of the PanoLunarTourbillon from Glashütte Original features an offset hour/minute display and tourbillon occupying the left-hand side of the dial, while the large twin date aperture is topped by an equally oversized moonphase indication.

> **Zenith**

El Primero Chronomaster Open Grande Date Moon & Sunphase

Using an ingenious system of superimposed discs, the automatic chronograph El Primero Chronomaster Open Grande Date Moon & Sunphase from Zenith displays not only the phases of the moon (at night), but also the course of the sun (during the day).

Patek Philippe, *Celestial Ref. 5102*

Equipped with a mobile map of the starry night sky, the highly complex Celestial from Patek Philippe provides a perpetual view of the exact configuration of the night sky over the Northern Hemisphere, complete with the visible movement of the stars and the moonphases, as well as the moon's position in the sky.

Chanel, *J12 Moonphase*

Made in ultra-resistant high-tech ceramic, the J12 Moon Phase by Chanel features a large aventurine subdial, swept over by a hand progressively pointing to the new moon, the first quarter, the full moon and the last quarter.

Louis Moinet, *Variograph*

On its Variograph model, Louis Moinet has opted to display, not the course of the moonphase, but the exact day of the full moon in the window at 6 o'clock.

Girard-Perregaux
Traveller Moon Phases and Large Date

Girard-Perregaux has equipped its Traveller Moon Phases and Large Date with an extremely evocative and accurate lunar display module that varies by only one day every 122 years. The instantaneous large date display moves at the stunning speed of 15/100ths of a second.

> **Piaget**, *Gouverneur G0A37115*

Piaget continues its grand tradition of ultra-thin expertise with this Gouverneur model combining a flying tourbillon with an "astronomical" moonphase (showing a one-day discrepancy every 122 years). The small hand precisely follows the waxing and waning phases of the heavenly body.

> **Graham**, *Geo.Graham The Moon*

The Geo.Graham The Moon from Graham combines a flying tourbillon with a retrograde moonphase display that needs to be corrected by only one day in 122 years. The openworked dial and the movement are set with 45 diamonds representing the constellations.

Hermès
Arceau Pocket Moonphase Retrograde Calendar

Powered by a movement from Vaucher Manufacture Fleurier, this Arceau pocket-watch from Hermès displays a large moonphase between 5 and 6 o'clock, as well as a retrograde date with central pointer. It is available in rose or white gold.

Ulysse Nardin, *Moonstruck*

A worthy heir to Ulysse Nardin's astronomical watches, the Moonstruck continuously indicates the position of the moon and sun relative to any point on the globe; it thus allows its wearer to read the current moonphase and the state of the tides wherever one may be.

Frédérique Constant
Slimline Moonphase Manufacture

Equipped with an in-house movement, the Slimline Moonphase Manufacture by Frédérique Constant is distinguished by its pure elegance and its entirely crown-adjustable moonphase display system.

Round Apertures

Alongside traditional apertures, we are also seeing an increasing number of round apertures, often equipped with mobile covers that represent the dark side of the moon. This enables horologers to create majestic extra-large and even extra-extra-large displays.

Jaquet Droz, *Eclipse Onyx*

On the Eclipse Onyx from Jaquet Droz, a small mobile black disc reproduces the moon's cycle by hiding and revealing a smiling moon, inspired by historical engravings. This spectacle is surrounded by eight applied stars.

Piaget, *Emperador Coussin Moonphase*

On Piaget's Emperador Coussin Moonphase, the disc representing the moon is a plaque of white gold that is heated to create craters such as those found on the face of the moon. Its hand-crafted production renders each watch one of a kind.

⌃ Bovet
Récital 9 Tourbillon Miss Alexandra

On the Récital 9 Tourbillon Miss Alexandra by Bovet, the lunar cycle is displayed by means of a fixed moon with an extremely realistic design that is progressively hidden or revealed by two dedicated discs reproducing the night sky.

⌃ Andersen Genève, *Kamar*

For its Kamar model ("kamar" means "moon" in Arabic), Andersen Genève has created a round aperture equipped with a mother-of-pearl disc bearing both a dark and a light moon, all on a dial of aventurine or guilloché 22-karat blue gold.

⌃ Martin Braun, *Selene*

On its Selene watch ("Selene" is the word for "moon" in ancient Greek), Martin Braun has opted to meld poetry and realism; a large moon based on a photograph is alternately hidden and revealed by a mobile cover.

⌃ MB&F, *MoonMachine*

MB&F gave free rein to the Finnish watchmaker Stepan Sarpaneva to create a special version of his HM3 Frog watch. This has resulted in the highly original MoonMachine, with a moonphase display via a round, crown-shaped aperture and a rotor that reproduces the constellations of the Northern Hemisphere.

Three-Dimensional Displays

A handful of watchmakers—most often independent artisans—have chosen to supplant the classic aperture systems with innovative three-dimensional displays. In these, the moon is depicted by a small mobile sphere (or a stationary sphere equipped with a mobile cover) reproducing the different faces of the moon throughout its cycle. This leads to very evocative stagings.

▼ De Bethune, *DB28 Skybridge*

The DB28 Skybridge by De Bethune combines a patented spherical moonphase with a night-sky dial in blued and mirror-polished grade-5 titanium, studded with white-gold and diamond stars.

▶ Cabestan, *Terra Luna*

The Terra Luna by Cabestan combines a three-dimensional moonphase indication featuring a Ø 7.4mm sphere with a vertical display on rotating drum barrels, along with a fusee-chain transmission system designed to compensate for the loss of force of the mainspring as it unwinds.

Konstantin Chaykin, *Lunokhod*

On the Lunokhod from Russian watchmaker Konstantin Chaykin, the heavenly body is represented by a stationary sphere sculpted in wootz steel, an Indian metal whose granular quality recalls the lunar surface. The moonphase is displayed by means of a semi-circular cover in black rhodiumed silver, which rotates around a small globe.

Cyrus, *Klepcys*

The Klepcys, designed by Cyrus in collaboration with the watchmaker Jean-François Mojon, boasts a black disc that slowly rises to hide a spherical moon, dotted with craters. During the new moon, the orb gives way to the brand's logo.

Christiaan van der Klaauw
Real Moon 1980

A three-dimensional moon takes pride of place on Christiaan van der Klaauw's Real Moon 1980. The astronomical indications are completed by a subdial that displays the height of the sun in relation to the horizon, as well as a hand at 3 o'clock signaling the eclipses of the sun and the moon.

Thomas Prescher
Mysterious Automatic Double Axis Tourbillon

On the Mysterious Automatic Double Axis Tourbillon from Thomas Prescher (whose movement is entirely hidden within the bezel), the two cylinders for the hours and minutes frame a spherical moonphase display.

The Feminine Side of the Moon

The moonphase display remains one of the favorite complications of ladies' watches, though it is still more widespread on masculine wrists. Catching the wave of this longstanding interest, both large brands and independent watchmakers offer a beautiful palette of quite refined ladies' models, most often adorned with diamonds and mother-of-pearl.

Breguet, *Reine de Naples*

Inspired by the wristwatch created for Napoleon's sister Caroline Murat, the Reine de Naples from Breguet has been introduced in a self-winding model with moonphase and a power reserve display.

Patek Philippe, *Diamond Ribbon Ref. 4968*

Illuminated by a diamond spiral that wraps around the case, Patek Philippe's Diamond Ribbon Ref. 4968 features a highly precise moonphase (accumulating a one-day discrepancy every 122 years) upon its mother-of-pearl dial.

Glashütte Original, *PanoMaticLunar*

Resolutely asymmetrical, the PanoMaticLunar from Glashütte Original displays the moonphase in a large vertical aperture at 2 o'clock, which also indicates the moon's "age"—above a large panoramic date characteristic of the German brand.

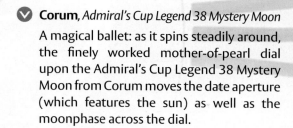

Saskia maaike Bouvier
The 8 Moons

Created by a woman for women, The 8 Moons timepiece from Saskia maaike Bouvier sports a starry night sky as a backdrop for eight spheres that show not only the current phase of the moon (at 12 o'clock), but also those for the seven following nights.

Corum, *Admiral's Cup Legend 38 Mystery Moon*

A magical ballet: as it spins steadily around, the finely worked mother-of-pearl dial upon the Admiral's Cup Legend 38 Mystery Moon from Corum moves the date aperture (which features the sun) as well as the moonphase across the dial.

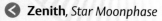

Zenith, *Star Moonphase*

The Star Moonphase from Zenith—equipped with the famous, ultra-thin, automatic Elite movement—features an imposing moonphase at 6 o'clock upon a dial punctuated by large blue Roman numerals. The cushion-shaped case is available in steel or rose gold.

Records for Precision

On ordinary moonphase watches, the complication will eventually show a one-day discrepancy after two years, seven months and around 20 days. On the moonphase complications within the luxury watches at the high end of the market, that difference has been reduced to one day over 122 years and 44 days. A few great brands have opted to face this technical challenge and develop moonphases that are much more precise—even resulting in no lag at all.

H. Moser & Cie
Moser Perpetual Moon

The Perpetual Moon from H. Moser & Cie would need to run 1027.3 years to differ from the real lunar cycle by one full day—thanks to a continually revolving display linked to the hours mechanism (as opposed to most moonphases, which advance once per day). The vertical lines marking the eight quarters of the moon, fixed by international standards, allow an ultra-precise reading.

A. Lange & Söhne, *1815 Moonphase*

The German company A. Lange & Söhne has equipped its 1815 Moonphase with a gear train fitted with a particular gear-train ratio that guarantees a difference of just 6.61 seconds per lunar month, or just one day every 1,058 years!

ochs und junior, *Moon Phase*

3478.27 years: with the Moon Phase model from the youthful ochs und junior brand, the famous watchmaker Ludwig Oechslin claims to have "created the world's most accurate moonphase wristwatch" while using just five components to modify the base movement. The dial also indicates the position of the moon in relation to the Earth and the sun.

Arnold & Son, *True Moon*

True Moon, from Arnold & Son, claims to provide a precise reproduction of the waxing and waning of the moon, adhering to the exact duration of synodic periods.

IWC, *Portuguese Perpetual Calendar*

Using an innovative system comprised of thee toothed wheels, the moonphase indication on IWC's Portuguese Perpetual Calendar has a discrepancy of just one day over 577 years. The double hemisphere display for the Northern and Southern Hemispheres is completed by two countdowns signaling the number of days until the next full moon.

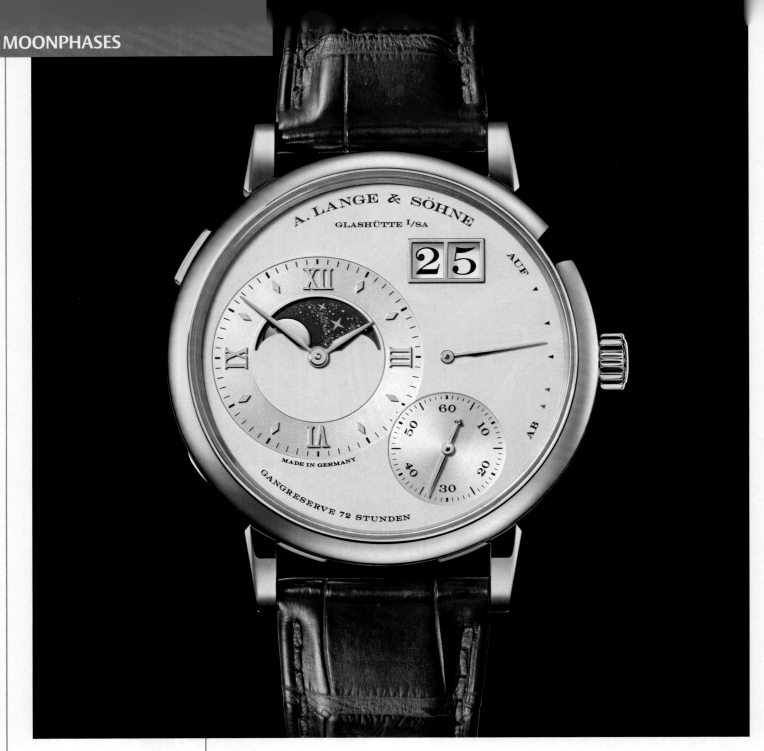

A. LANGE & SÖHNE

GRAND LANGE 1 MOON PHASE – REF. 139.032

Positioned prominently within the watch's principal subdial at 9:00, the hand-wound L095.3 caliber's moonphase takes on the lunar cycle with bold ingenuity and distinguished sophistication. The intricately coated night sky, adorned with more than 300 stars of varying sizes, depicts a vivid moon that, thanks to its connection to the caliber's hour-wheel continuum, is in constant motion through its monthly journey. In addition to this touch of undetectable, yet extraordinary, mechanical detail, the timepiece is endowed with a seven-gear transmission that reduces the lunar cycle's deviation to less than one minute, requiring a mere one-day correction every 122.6 years. The 41mm pink-gold wristwatch is finished with a large date, subsidiary seconds and a 3:00 indicator of the movement's 72-hour power reserve.

BREGUET

CLASSIQUE MOON PHASES – REF. 7787BR.29.9V6

Radiating Breguet's signature style of classical elegance, this 39mm 18K red-gold timepiece boasts an exquisite hand-engraved dial with two perfectly positioned indications. Joining the off-centered power reserve scale on the lower half of the watch, a superb moonphase at 12:00 enlightens the wearer on both the age and phase of the lunar cycle. Meticulously finished with a hand-engraved 18K gold oscillating weight, the 591 DRL caliber is enhanced with a Breguet balance wheel and silicon straight-line lever escapement.

BREGUET

CLASSIQUE HORA MUNDI – REF. 5717BR.US.9ZU

A world-first in mechanical watchmaking, the Classique Hora Mundi's self-winding 77FO caliber takes the watch's wearer on an unparalleled voyage in timekeeping. Equipped with a groundbreaking synchronized module with reprogrammable mechanical memory wheel, the timepiece travels from one pre-selected city to the next with a single push of the crown at 8:00. Exquisitely embellished with a variety of refined techniques, the entirety of the globe-motif dial's indications—moonphase, day/night indicator, date and time via two open-tipped blued steel Breguet hands—instantly jumps to the specifications of the chosen destination.

BREGUET

HERITAGE PHASE DE LUNE RETROGRADE – REF. 8860BR.11.386

Breguet has added a new model to the Heritage collection, equipped with a moonphase. The automatic-winding caliber 586L movement indicates the hours and minutes in the dial center of engine-turned mother of pearl, while a moon matching the case material occupies an aperture at 12:30. An additional specially shaped plate was made for the retrograde moonphase indication without adding to the height of the movement. Attention to detail is a hallmark of this piece; although invisible to the naked eye, the face on the moon is actually winking.

BREGUET

REINE DE NAPLES JOUR/NUIT – REF. 8998BB.11.874

Breguet expresses the grace of the dance between the day and night with the Reine de Naples Jour/Nuit. The manufacture used a disc of lapis lazuli for the sky and adorned it with clouds of white mother-of-pearl, gold stars and a moon of engraved titanium. The true star on this model is the sun, symbolized by the facetted rim of the balance wheel that exudes rays of light. While the balance does its duty, the sun goes on a daylong tour, passing beneath the steel bridge supporting the mechanism, meeting the hands for the hours and minutes and reaching its zenith facing the titanium moon.

CHANEL

J12 MOONPHASE – REF. H3407

A watchmaking icon of the 21st century, the J12 watch has reinvented itself season after season while still respecting the essence of its design. A heavenly body illuminated by the sun and a symbol of femininity, the moon has been veiled in mystery since the dawn of time. The elegant silhouette in high-tech ceramic reveals, in an aventurine starry sky, the phases of the moon that reveal their secrets to those who know how to observe them.
The J12 Moonphase is also available in white and black high-tech ceramic classic or set with diamonds.

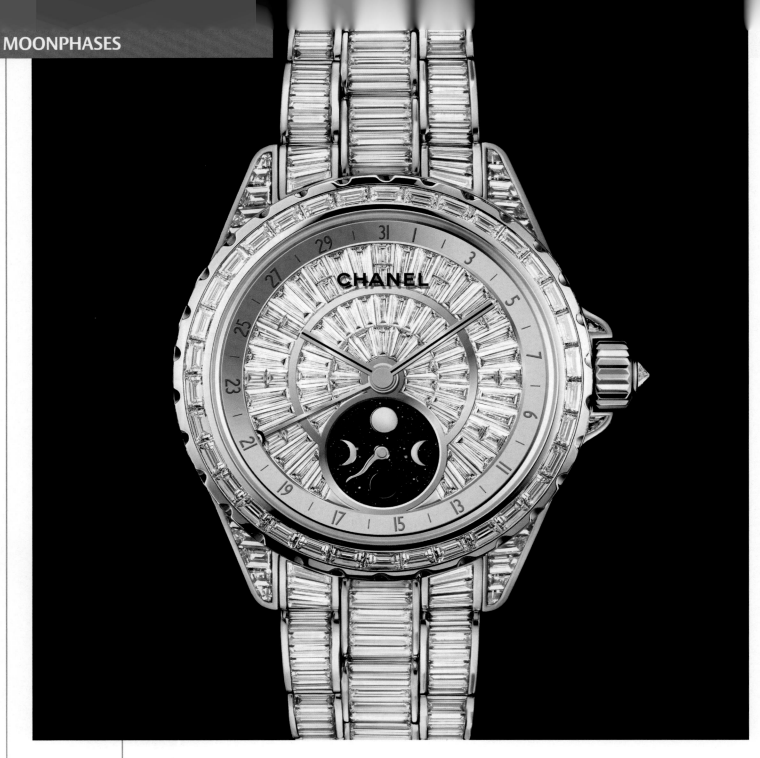

CHANEL

J12 MOONPHASE FULL SET – REF. H3426

Blending modernity, elegance and purity, the J12 watch offers a new version of its shine: a moonphase watch. Unlike classic moonphase watches where the phases appear through an aperture, those of the J12 Moonphase watch are displayed on a deep-blue aventurine disc sparkling like a summer's night. Positioned at 6:00, the disc features the four main phases of the moon and is adorned with a little serpentine hand.

A subtle distinction, the date is indicated by a central hand poetically tipped with the shape of a crescent moon. This piece is part of a limited edition of 5 pieces in 18K White Gold and 696 baguette-cut diamonds (approx. 42.45 carats).

DE BETHUNE

DB25L JEWELLERY

This creation is distinguished by the many ways in which the light plays across the materials and the complex dial architecture. The dark gleam of hand-polished and blued titanium, the sparkling twinkle of the white gold and diamond stars, and the muted glow of the curved blued steel hands make for a stunning piece. The depth of this star-studded sky, which can be personalized with date and location is revealed through the alternating curves that are convex on the outside for the chapter ring, and concave in the center. A spherical moon set with sapphires and diamonds shines at 12:00, providing an exceptionally accurate moonphase display that will diverge from astronomical reality by just one day every 122 years.

DE BETHUNE

DB25LWS3V2

Powered by a manual-winding DB 2105 V2 caliber, this model is equipped with 27 jewels and a power reserve of 6 days, and beats at 28,800 vph. This exceptional timepiece is housed in a shining white-gold drum-shaped case with cone-shaped lugs and a crown at 3:00. The bold blue hue of the De Bethune star-studded sky on the dial, which can be personalized, sets the perfect display for the platinum and flame-blued steel spherical moonphase indicator at 12:00. This watch is finished with an extra-supple alligator leather strap with a pin buckle.

DE BETHUNE

DB28 BLACK MATTE

The DB28 won the "Aiguille d'Or" in 2011, the most honorable distinction offered by the Geneva Watchmaking Grand Prix. This limited edition created of this iconic model has its distinctly shaped case in sandblasted anthracite zirconium is both understated and bold, forming the perfect frame for the intricate dial. A black mirror-polished and silver-toned minutes ring surrounds a mirror-polished steel bridge and the exclusive Côtes De Bethune motif. The spherical moonphase at 6:00 is crafted in platinum and polished anthracite.

DE BETHUNE

DB28 SKYBRIDGE

Ancestral skills and the latest scientific breakthroughs merge to make this timepiece a perfect summary of highly aesthetic ambition and peerless technical precision. While classically inspired in terms of its round shape, its crown at 12:00 and its hunter-type back borrowed from pocket-watches, the ultra-light mirror-polished titanium case of the DB28 with its distinctive floating lugs nonetheless proclaims the collection's modern and indeed futuristic nature. An eminent reference to 18th-century clocks, the star-studded sky in mirror-polished and blued titanium derives its power and beauty from its concave shape and multiple decorations. The spherical moonphase display is accentuated by an arrow-shaped bridge that appears to be pointing towards infinity.

de GRISOGONO

OTTURATORE – REF. OTTURATORE N03

A hyper-efficient module permits this self-winding timepiece's inner dial to rotate 90° in less than two hundredths of a second, or 15 times the speed of the blink of an eye. Visibly undetectable (thanks to its extraordinary control of powerful counter forces and ability to achieve instantaneous maximum velocity and an immediate stop), the dial, decorated with Clou de Paris, revolves to position its lone aperture above one of the watch's four individual complications—moonphase, date, power reserve and small seconds. The Otturatore is housed in a pink-gold case.

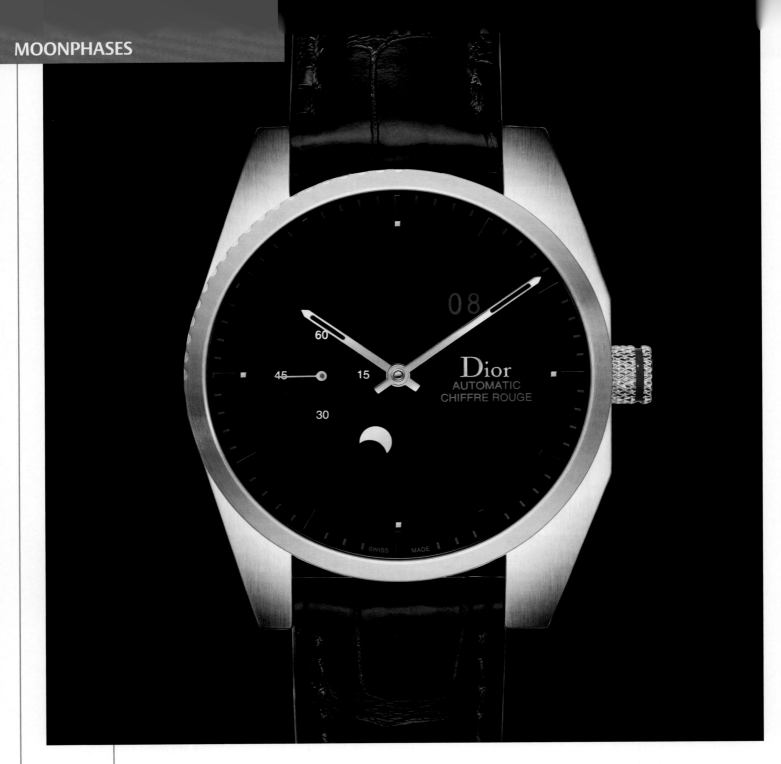

DIOR

CHIFFRE ROUGE C03 – REF. CD084C11A001

The Chiffre Rouge collection is enriched by a new moonphase watch, equipped with an Elite movement made by Zenith manufacture. Reinterpreting facets of the Dior Homme universe and its codes, the moonphase indicator at 6:00 and the small second counter at 9:00 play with contrasts of matte and shiny black while the large date is displayed in a bold red at 2:00. This elegant timepiece, with a sleek graphic design, is available in a limited edition of 100 pieces.

F.P. JOURNE

OCTA LUNE

The Octa collection caters to a clientele looking for authentic, rare and innovative horology in an "easy-to-wear" mechanical wristwatch. The automatic-winding Octa Lune offers over 120 hours of chronometric power reserve, reminding us of ancient times when the full moon was the only source of light during the night. While today we no longer depend on moonlight for night vision, the moonphase indication remains one of the most poetic of all horological complications. The hours, placed on the right side of the dial, are easily and quickly read. The Octa Lune won "Men's Watch of the Year" at the 2003 Grand Prix d'Horlogerie in Geneva.

FRANCK MULLER

MASTER BANKER LUNAR – REF. 8880 MB L WGE

Franck Muller introduced the Master Banker line for businessmen who are constantly on the go. This model offers three different time zones that are adjusted via the crown so the wearer easily knows the time at the stock markets of his choosing around the world. The day/night display of the two time zones is located in the upper part of the dial next to the counters with a small conjuring trick to ensure clear legibility beyond comparison. The Lunar model is also complete with a large and artful moonphase indicator at 6:00.

GLASHÜTTE ORIGINAL

PANOMATICLUNAR – REF. 90-02-42-32-05

The Pano collection's signature asymmetric effect is easily appreciated with a glance at the new PanoMaticLunar, which features the characteristic off-center visuals. The hour/minute and small seconds subdials are aligned along a vertical axis to the left, while the Panorama Date and superb moonphase display are positioned to the lower and upper right, respectively. The stainless steel case frames a warm silver dial with silver hour indexes beneath blued steel hands, complemented by the moonphase display and characteristic Panorama Date. Within beats the manufacture's superb Calibre 90-02, a self-winding 47-jewel movement with a power reserve of 42 hours. The sapphire crystal caseback provides a clear view of the finely finished movement, including the typical Glashütte three-quarter plate with Glashütte ribbing, off-center skeletonized 21K gold rotor, blued screws and hand-engraved balance cock.

PATEK PHILIPPE

CALATRAVA – REF. 7121

Worn on a pearly beige hand-stitched alligator strap, this seductive ladies' timepiece flaunts its golden personality. Housed in a 33mm 18K yellow-gold case illuminated by the shimmer of 66 flawless Top Wesselton diamonds, the ultra-thin hand-wound 215 PS LU caliber boasts a highly precise moonphase complication that needs only one adjustment every 122 years. Combined with a hand-guided small seconds display for an optimal use of space, the golden moon is exhibited upon a starry sky in a subdial at 6:00. The hours and minutes are indicated via two "Poire Stuart" hands in 18K gold, against 11 Breguet numerals in the same precious metal, all on a cream-colored brass dial with a delicately grained texture.

PATEK PHILIPPE

MEN COMPLICATIONS – REF. 5396/1R-010

This 38.5mm 18K rose-gold timepiece with radiant rose-gold bracelet masters the art of architectural efficiency, conducting a highly sophisticated mechanical choreography with impressive understated elegance. Displaying the hours, minutes and seconds via three central rose-gold hands, day and month via side-by-side apertures at 12:00, moonphase in a colorful 6:00 subdial framed by a 24-hour indicator and date in a discreet window at the bottom of the silvery opaline dial, this annual calendar is a spectacle of structural efficiency. Its 324 S QA LU 24H caliber is composed of 347 parts, oscillates at a frequency of 28,800 vph, and boasts a power reserve of 35-45 hours.

Diving

Hublot
Oceanograph
4000M

Georges Kern
CEO of IWC

"The brand is increasingly high-end, and the new Aquatimer collection expresses that evolution perfectly."

"MORE MANUFACTURE-MADE MOVEMENTS, NEW MATERIALS AND UNPRECEDENTED TECHNICAL SPECIFICATIONS." In a few words, that is how Georges Kern, CEO of IWC, describes the new Aquatimer Collection from the Schaffhausen manufacture.

What does 2014 hold for IWC?
For one, we will continue to modernize our products, as we have always done. Our watches have evolved greatly, rising not only in perceived value, but also in actual value. This is largely due to the fact that we are producing more and more of our own calibers, and in-house movements now represent the majority of our sales. IWC is an engineering brand; our work consists of being technically innovative and developing movements, modules and functions for the consumer. We invest a lot of money in this field.

What guides your distribution strategy?
We follow a policy of selective and qualitative distribution. At this point, all the big watchmaking brands, following the example of jewelry brands, have their own stores across the globe. IWC currently has about 60 stores. And not just in China! We've opened locations in Europe, including Paris, Zurich and Rome, but also in the United States, for example in California.

The year 2014 is dedicated to the diving watch at IWC. Indeed, a completely reimagined Aquatimer collection was the big news for the brand at the beginning of the year. How would you describe this launch in a few words?
The brand is becoming increasingly high-end as it matures. The new Aquatimer collection expresses that evolution

perfectly. We are incorporating more in-house movements, new materials and unprecedented technical specifications.

Where does this collection fit in among IWC's other collections?
We have three lines that I would describe as "elegant"— Portuguese, Portofino and Da Vinci—and three "sporty" lines—Aquatimer, Ingenieur and Pilot—for a perfect balance. Naturally, we want to maintain this elegant, sporty image. In fact, each of our collections plays a role in achieving this equilibrium.

> Aquatimer Perpetual Calendar Digital Date and Month

⌄ Aquatimer Deep Three

Georges Kern, CEO of IWC

> Aquatimer Automatic 2000

The market for diving watches is a niche one, with relatively small sales. So why is it so important?

It's true that the diving watch is a niche. But it's an important one, because the diving watch is the ideal object with which to demonstrate our technical expertise. At IWC, we are known for our technically sophisticated watches and the Aquatimer collection has largely contributed to this image—and will continue to do so. In addition, the world of the diving watch has been in IWC's DNA since 1967. It is a true signature for the brand.

The brand has matured; would you say the same is true of the Aquatimer?

I really think so. And one look at the products in the collection is enough to remove any doubt. The collection is a little more restrained and muted than previous Aquatimer models. These watches are still true technical instruments, but they are now also elegant instruments. The vivid colors—yellow or orange—that characterized Aquatimer models in the past are now gone. All of that speaks to a greater maturity. IWC no longer needs these distinctive signs to make its collections stand out. Our current collections are different from the preceding ones both in real and perceived value, especially on a technical level.

On that note, one of the collection's major innovations lies in the rotating bezel, a new IWC patent. How would you describe it?

Since 1967 and the first Aquatimer, IWC diving watches have been characterized by the use of an internal rotating bezel. It became a brand icon. Although it had the advantage of being protected from exterior shocks and other damage, it was not necessarily easy to handle, especially during a dive. That is why IWC reconsidered the more conventional exterior rotating bezels when we last revisited the collection in 2009. This year, the collection's new models all boast a double interior/exterior rotating bezel, which possesses all of the advantages of both systems, with none of the drawbacks. The new SafeDive system combines the benefits of an internal rotating bezel—notably the protection of the mechanism against salt water and dirt, and the precise adjustment of dive stages by the minute—with the ease of use provided by an external rotating bezel, which one can adjust easily even with diving gloves or cold hands. In short, with the double rotating bezel, we have achieved the best of both worlds.

Another of IWC's patents brings an additional asset to this new collection...

Yes, another key element that ties together all nine models in the collection is our patented system of interchangeable straps. The new system is not only practical—the change from stainless steel bracelet to rubber strap and vice versa is quick and easy—but also very safe. The bracelet is attached to the locking bar from above and engages with an audible click. To release it, the locking lever is pressed outward with the thumb and the bracelet pushed upward. In its XXL version, the corrugated rubber strap can even be worn over a drysuit. These are important elements for a diving watch, since some divers—especially those who wear the watch over their suit—change the strap frequently, and if that change is too complicated, the diver will give up on it.

That leads us to something else that you have often underscored, which is that IWC makes watches, not just movements.

That's actually essential, because the client buys an entire watch, not just a movement. At IWC, our R&D teams work not only on the movement, but also the rest of the watch. That is how we were able to resolve important problems such as the double rotating bezel—which offers, may I remind you, the best of both worlds in terms of the internal/external rotating bezel—and the interchangeable strap.

If you had to choose one emblematic model from the new collection, which would it be?

It would be impossible, and much too simplistic, to choose just one. But I will say that among the new models, I would highlight the Aquatimer Perpetual Calendar Digital Date and Month which, besides integrating an haute horology caliber within a very sporty watch, boasts a 49mm diameter, making it the second-biggest watch in the history of IWC! And I can't forget to mention the Aquatimer Deep Three, which is the latest member of a beautiful family. After the Deep One model in 1999 and the Deep Two model in 2009, the Aquatimer Deep Three presents, like Deep One and Two, a mechanical depth gauge, but with improvements: the blue pointer indicates the current depth, while the red pointer stays at the maximum depth achieved.

And for those who go deep diving…

For that kind of extreme athlete, we offer the Aquatimer Automatic 2000, which has a water resistance to 200 bar, practically equivalent to 2,000m. Its titanium case is a reminder that IWC was the first brand to use this material in a series in 1982, with the Ocean 2000, designed by Ferdinand A. Porsche.

IWC is also releasing a diving watch with a chronograph this year.

Yes, our range of chronographs is very rich and varied this year with the new Aquatimer collection. I will mention just three, all of which are equipped with an in-house movement, the 89365 caliber: the Aquatimer Chronograph Edition "50 Years Science for Galapagos," the Aquatimer Chronograph Edition "Galapagos Islands," both in steel covered in black rubber, and the Aquatimer Chronograph Edition "Expedition Charles Darwin," which is notable for its bronze case, a first for IWC.

> Aquatimer Chronograph Edition "Galapagos Islands"

> Aquatimer Chronograph Edition "Expedition Charles Darwin"

> Aquatimer Chronograph Edition "50 Years Science for Galapagos"

Diving Watches

With their sturdy cases designed to withstand the pressures exercised below the world's seas and oceans and their dials that are ultra-legible even at great depths, as well as their resolutely technical look, diver's watches have asserted themselves as the quintessential sports models. That is undeniably why they appeal not only to professional divers and authentic extreme adventurers, but also to all enthusiasts of sporting feats wishing to share something of these thrills. From models water resistant to 100m to those able to face the Mariana Trench, we are seeing a wide range of diver's watches equipped with various operating systems and safety devices. The last few years have witnessed a growing number of instruments equipped with mechanical depth gauges as well as models with audible alarms (see the chapter on alarm watches). Designers are increasingly attentive to aesthetic and style aspects, offering chic diver's models that are equally at home in daily life as in the solitude reigning beneath the ocean.

WATER RESISTANCE, AS WELL AS LUBRICATION, HAS ALWAYS CONSTITUTED ONE OF HOROLOGERS' MAIN CONCERNS.

This is not to say that all horological aficionados are budding Jacques Cousteaus. However, water resistance represents a welcome protection, even on terra firma, against the two great enemies of watch movements: humidity and dust.

The first water resistant watch, the famous Oyster from Rolex, arrived on the scene in 1926, thanks to the invention of the "Screw Crown." In 1953, the brand created for Professor Auguste Piccard a watch that would accompany his bathyscaphe to 3,150m under the sea. In 1960, the Deep Sea Special would descend to over 10,916m down in the Marianna Trench in the Pacific Ocean, and ascended to the surface intact after having undergone pressure of more than a ton per square centimeter.

Besides explorers and scientists, the first diving watches often equipped the armed forces. This was notably the case with Panerai in the 1930s and 1940s, made for the Italian Navy, with the models' luminescent indications on the dial and a crown system guaranteeing water resistance to 200m.

In the 1950s, with the rise of water sports, came the first collections devoted to diving watches. Rolex set the tone in 1953 with its famous Submariner, water resistant to 100m (now 300m). That same year, Blancpain launched its Fifty Fathoms, created in collaboration with two "frogmen" from the French Army. This watch, initially resistant to 50 fathoms of depths—hence its name—stands out for its black dial, bezel and strap. It would later be released in countless other versions.

Ever since, diving watches have become the quintessential sporting watch, and horologers have continued to increase their resistance and robustness

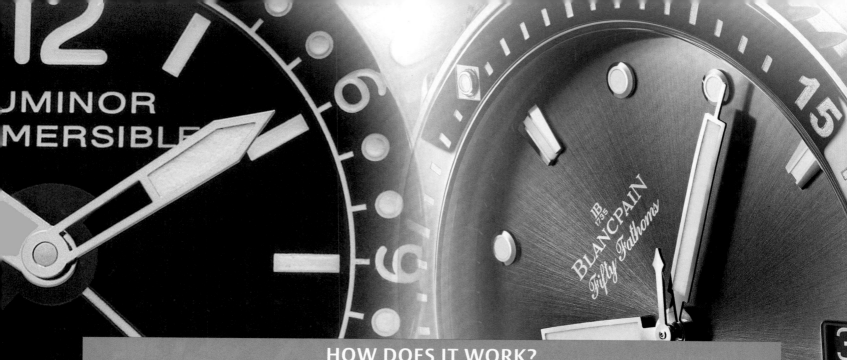

HOW DOES IT WORK?

The definition of an authentic "diving watch" conforms to an official Swiss industry standard, or "norme de l'industrie horlogère suisse" (NIHS 92-11). To earn the title of diving watch, a timepiece must not just possess a water resistance superior to the average watch (minimum 100m), usually ensured by a screw-down crown and caseback. It must also offer all the functions that are indispensable to amateur or professional divers, notably a system that allows one to preset and check the time of the dive, with a device preventing any accidental manipulation. This is usually done by a turning bezel whose unidirectional rotation prevents the diver from inadvertently extending the remaining dive time. The dial usually features large numerals and highly visible hands, generously covered in luminescent coating. The metal strap or rubber bracelet is often equipped with an extension system that stretches over the diving suit. The highest-performance models possess a valve that eliminates any helium within the watch. This highly volatile gas, used to facilitate breathing in pressurized chambers, gets inside the watch case very easily, but the case must be totally evacuated of helium when re-ascending to avoid the build-up of excessive pressure.

A SMALL REMINDER ABOUT WATER RESISTANCE

Contrary to what is all too often believed, the indication of water resistance in meters is a technical standard that does not correspond to an absolute value. The phrase "water resistant to 30m" does not imply that the watch can be worn to that depth under any conditions whatsoever. The movements of the wearer (dives, jumps, swimming, etc.) can increase the pressure considerably. For all aquatic activities, it is therefore best to stick to watches that possess a water resistance of at least 100m.

DIVING CHRONOGRAPHS

The inherently sporty nature of the two functions means that horologers often add chronographs to diving watches. However, it bears noting that most of these instruments, though perfectly "water resistant," cannot be activated in chronograph mode during the dive, or else water will seep into the case via the pushbuttons.

Re-Releasing Flagship Models

Several well-known brands are re-releasing or re-interpreting their emblematic diving watches, especially the pioneering models of the 1950s. Though their look might be similar to the original, their performances usually benefit from the enormous technical progress made in the intervening years.

Vulcain, *Nautical Heritage Seventies*

A tribute to a 1970s model, the Vulcain Nautical Heritage Seventies is distinguished by an alarm that is easily audible under water, water resistance to 300m and a decompression-stop indication system. This 300-piece limited edition sports a resolutely vintage look.

Rolex, *Oyster Perpetual Submariner*

Heir to the legendary 1953 model, the Rolex Oyster Perpetual Submariner is water resistant to 300m thanks to a patented Triplock screw-down crown system. The unidirectional rotating bezel is in ultra-resistant scratchproof ceramics.

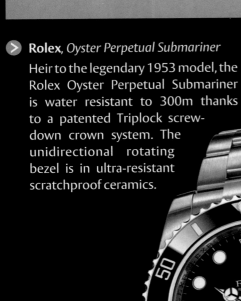

Omega
Seamaster PloProf 1200M

Omega re-interpreted its 1970 Seamaster 600M model, also called PloProf (for "plongeurs professionnels," or professional divers), in a modernized version that is water resistant to 1200m, with a co-axial escapement and a security pushbutton that locks and unlocks the rotating bezel.

Tudor, *Heritage Black Bay*

Inspired by a 1954 model, Tudor's Heritage Black Bay model (water resistant to 200m) stands out for its "vintage chic" design, with a convex dial, red rotating bezel, "snowflake" hands and large luminescent hour markers.

Blancpain, *Fifty Fathoms Bathyscaphe*

To mark the 60th anniversary of its historical Fifty Fathoms model (1953), which has now given rise to an entire collection of diver's watches and chronographs, Blancpain has created the Fifty Fathoms Bathyscaphe, inspired by the submarine of the Swiss adventurer Jacques Piccard.

Champions of Style

From sporty chic models destined mainly for poolside cocktails to professional instruments equipped for confronting the watery depths, from ultra-technical looks to timepieces focused on class and elegance, diving watches exist in a vast array of choices. Steel occasionally gives way to unusual materials—ceramics, titanium, bronze—and the quest for high performance is increasingly paired with close attention to style. These are watches that shine underwater, but also on dry land.

> **Ulysse Nardin**, *Marine Diver Black Sea*

Water resistant to 200m, the Marine Diver Black Sea by Ulysse Nardin stands out with its total-black look offset by blue hands and hour markers, as well as its steel case clad in vulcanized rubber. The dial and the unidirectional bezel are adorned with a wave pattern.

> **Omega**

Seamaster Planet Ocean 600M GMT GoodPlanet

Created in honor of the GoodPlanet Foundation, which is dedicated to preserving the oceans, the Seamaster Planet Ocean 600M GMT GoodPlanet combines a movement featuring a co-axial escapement with a system that displays three time zones.

> **Maurice Lacroix**, *Pontos S Extreme*

On the Pontos S Extreme by Maurice Lacroix, crafted in an ultra-light alloy named Powerlite®, dive-time control is handled via a crown-driven rotating inner bezel ring.

Gucci, *Gucci Dive*

Powered by a Girard-Perregaux movement, the Gucci Dive watch (water resistant to 300m) comes in steel or pink-gold versions, with a black PVD coating on both models. It features an eminently readable and elegant design.

TechnoMarine, *Black Reef*

The Black Reef by TechnoMarine teams a titanium case (water resistant to 500m) with a mesh bracelet. It is equipped with a valve designed to vent helium as well as oversized hour markers appearing on a wave-motif dial base.

Audemars Piguet
Royal Oak Offshore Diver 42 mm

The Royal Oak Offshore Diver 42 mm by Audemars Piguet has been interpreted in a stylish matte black high-tech ceramic version. The rotating inner bezel ring, which calculates dive times, is adjusted via a crown at 10 o'clock.

Super Divers

Certain companies have chosen to compete by developing exceptional instruments capable of descending to 1,000m, 3,000m, 6,000m or even 12,000m underwater! These super divers possess ultra-robust and innovative architectures, reinforced water resistance systems and even— for the highest performers—cases filled with liquid that is immune to pressure.

Breitling, *Superocean Chronograph M2000*

With a patented system of magnetic pushbuttons that operate through the case itself, the Superocean Chronograph M2000 from Breitling is presented as the only chronograph in the world that is watertight and functional at 2,000m underwater.

Girard-Perregaux, *Seahawk II Pro*

On its Seahawk II Pro, water resistant to 3,000m, Girard-Perregaux has shifted the crown to 4 o'clock to prevent it from getting in the way of the movements of the wrist, and equipped it with a large crown-guard that is both ergonomic and highly original.

CX Swiss Military Watch™
20 000 FEET

The watchmaker CX Swiss Military Watch™ struck a blow for the record books with its 20 000 FEET, a mechanical timepiece that can plunge to more than 6,000m of depth.

Bell & Ross, *Hydromax*

Equipped with a quartz movement, the Hydromax from Bell & Ross is water resistant to the amazing depth of 11,000m, thanks to its case, filled with an oily liquid that is immune to pressure.

Doxa, *SUB 4000T Professional*

Water resistant to 4,000 feet (1,220m), Doxa's SUB 4000T Professional stands out by its orange dial, helium valve, "Safe Dive" power reserve indicator and its asymmetrical case, shaped to protect the crown.

Porsche Design, *P'6780 Diver*

On the P'6780 Diver from Porsche Design, water resistant to 1,000m, the automatic movement is lodged inside a pivoting chamber. When the chamber is locked, the rotating bezel and crown are immobilized. The user must unlock it to set

Pita, *Oceana*

Using two patented mechanisms that enable time setting by rotating the base and "reinforced magnetic gears," the Oceana model from Barcelonan brand Pita—water resistant to 5,000m or 2,000m—can lay claim to being "the only watch that you can set underwater."

Rolex
Oyster Perpetual Deepsea Challenge

Water resistant to 12,000m, Rolex's experimental diving watch Oyster Perpetual Deepsea Challenge accompanied director and explorer James Cameron in 2012 to 10,898m underwater in the Mariana Trench, the deepest solo dive in history.

Rolex, *Sea-Dweller Deepsea*

Rolex has reaffirmed its sovereignty over the deep seas with its launch of the Sea-Dweller Deepsea, water resistant to 3,900m. To resist such colossal pressures, this extraordinary instrument benefits from a patented new case architecture called Ringlock.

Super Luminous

Diving watches must offer optimal readability in the extreme darkness prevailing at great depths. For this purpose, they are generally equipped with oversized numerals, indexes and hands that are treated with luminescent coating. Some brands have developed "lighting" systems offering even higher performance.

Ball Watch, *Engineer Hydrocarbon NEDU*

On Ball Watch's Engineer Hydrocarbon NEDU (water resistant to 600m), the hour markers and hands are equipped with micro-tubes filled with a luminescent gas. This Swiss technology guarantees a luminosity 100 times more intense than other methods.

Luminox, *Deep Dive*

The Luminox Light Technology System, in operation on the Deep Dive by Luminox, guarantees visibility in any conditions, thanks to its micro-tubes reflecting a blue light—the last color perceivable by the human eye after a certain depth.

Watches with Depth Gauges

Over the last few years, we have witnessed the return of a function that is not strictly horological, but very useful for divers: the depth gauge, or bathometer. It is a tempting alternative—or complement—to the electronic wristworn computers that are currently flooding the market.

Jaeger-LeCoultre, *Master Compressor Diving Pro Geographic Navy SEALs*

The Master Compressor Diving Pro Geographic Navy SEALs from Jaeger-LeCoultre stands out for its mechanical depth gauge equipped with a membrane that dilates or contracts according to the water pressure and then acts upon a large arrow-shaped hand. The depths measured range from 0 to 80m.

IWC, *Aquatimer Deep Two*

The Aquatimer Deep Two from IWC is equipped with a mechanical depth gauge with two differently colored hands, one indicating the current depth (up to 60m), and the other the maximum depth achieved during the dive—a system that guarantees maximum safety.

Oris, *Aquis Depth Gauge*

The Aquis Depth Gauge by Oris, which is water resistant to 500m, enables the water to seep into a semi-circular gauge positioned beneath the sapphire crystal. When the pressure rises, the volume of the gas contracts, and the depth is indicated by the point of contact between the water and the gas.

Panerai, *Luminor 1950 Submersible Depth Gauge*

On its Luminor 1950 Submersible Depth Gauge, Panerai pairs an automatic caliber with an electronic depth gauge using a membrane and microchip system that activates a large yellow central hand. The levered bridge protecting the crown ensures water resistance to 120m.

Blancpain, *X Fathoms*

Blancpain's X Fathoms houses a mechanical depth gauge that operates to 90m, with memory of the maximum depth achieved and a separate, ultra-precise (+/- 30cm) display for the 0-15m zone, as well as a five-minute retrograde counter for the decompression stages.

AUDEMARS PIGUET

ROYAL OAK OFFSHORE DIVER – REF. 15706AU.OO.A002CA.01

The Royal Oak Offshore Diver is powered by Audemars Piguet's in-house, automatic-winding Manufacture caliber 3120 and is 42mm in diameter. The movement, comprised of 280 parts and beating at 21,600 vph, is enclosed in a forged carbon case with a black ceramic bezel, black rubber-molded crown and pushbutton. The dial features a rotating inner bezel ring bearing a graduated diving scale and luminescent coating on hands and hour markers. Its functions include dive-time measurement, hours, minutes, seconds and date. The case, water resistant to 300m, is mounted on a rubber strap with oversized titanium buckle.

BLANCPAIN

X FATHOMS – REF. 5018-1230-64

The X Fathoms revisits the characteristics of its iconic 1953 ancestor, combining them with a mechanical depth gauge to create the highest-performance mechanical diving watch ever produced. Depth measure up to 90m and maximum depth reached memory, separate indication on the 0-15m scale with exceptional +/- 30cm precision, retrograde five-minute counter for decompression stops—the X Fathoms concept watch abounds in world firsts. Its movement, reference 9918B, is based on the manufacture-made Calibre 1315, which powers several models in the Fifty Fathoms collection. Its imposing 55.65mm case made of satin-brushed titanium is water resistant to 300m. It features a helium decompression valve for saturation diving and the unidirectional rotating bezel characteristic to the collection for almost 60 years. The individual calibration of each X Fathoms watch guarantees maximal precision of depth indications.

BREGUET

MARINE ROYALE – REF. 5847BR.Z2.5ZV

Capitalizing on water's superior sound conduction, carrying tones approximately four times faster than above the surface, this 45mm 18K rose-gold timepiece brings sonorous enchantment to the world of diver's watches. An authentic diving instrument first and foremost, the self-winding timepiece boasts a unidirectional rotating bezel, as well as luminously enhanced 18K gold hands, optimized for legibility in murky, submerged conditions. Set at 4:00, the alarm's status is discreetly clarified via an aperture at 12:00 on the black rhodium-finished 18K gold dial with rose-engine engraved wave pattern. Armed with a 36-hour power reserve, the 519R caliber boasts an 18K rose-gold rotor.

HUBLOT

OCEANOGRAPHIC 4000M – REF. 731.NX.1190.RX

The product of 18 months of determined research, development and testing, the Oceanographic 4000M is nothing short of a groundbreaking watchmaking achievement. Tested to the depth of 5000m, this self-winding diving masterpiece is certified to an astonishing 4000m below the surface. Accounting for the extreme pressure of such depths, the synthetic sapphire crystal requires an extraordinary thickness to ensure watertightness. From a legibility standpoint, the 48mm titanium timepiece's dial components are optimized in size and treated with SuperLumiNova for underwater luminescence at points of intense darkness. Finished with a helium valve for the release of gases during ascent, the Oceanographic 4000M is worn on a rubber and nylon blended strap, sized longer than ordinary to be worn over a diver's protective suit.

AQUATIMER CHRONOGRAPH – REF. IW3769

The red-gold version of the Aquatimer Chronograph demonstrates once again that elegance is not at variance with a highly functional sports watch at IWC Schaffhausen. Allowing the wearer to change directly from a wetsuit to a dinner jacket, this versatile model's 44mm case guarantees that even the most stylish diver can plunge into the great depths of the sea. The rotating diving bezel is unlike any before, as it features "background lighting." Pressed into the multi-component bezel is a 4mm-wide sapphire crystal ring, to which a concentrated coating of SuperLumiNova is applied to its underside. The system is thus particularly suited to night dives, making this watch extremely functional.

AQUATIMER CHRONOGRAPH EDITION GALAPAGOS ISLANDS – REF. IW3767

The Aquatimer Chronograph Edition Galapagos Islands not only represents the harmonious development of IWC's of professional diver's watches, but is also the star attraction to mark a new partnership by IWC Schaffhausen and the Charles Darwin Foundation in aid of the environment, particularly on the Galapagos Islands. This model is an attractive, useful and extremely sporting symbol of this collaboration: black like the lava of the volcanic islands born from the ocean, and white like the clouds floating by above them. An intricate relief engraving of a giant tortoise on the back of the case, surrounded by the inscription "Tribute to the Charles Darwin Foundation – Galapagos Islands," represents the Charles Darwin Foundation and the Galapagos Islands.

PANERAI

LUMINOR SUBMERSIBLE 1950 2500m 3 DAYS AUTOMATIC TITANIO – 47MM
REF. PAM00364

This special edition model is produced in a limited series of 500 pieces. The brushed titanium unidirectional bezel on this 47mm timepiece rotates counterclockwise with a graduated scale for calculating the time of immersion, as well as ratchet click at minute intervals. Completing the timepiece is a black rubber strap, fastened by a large trapezoid-shaped buckle in brushed titanium. The strap can be changed quite easily by means of the button fitted into the lug of the watch, which is operated by a special tool provided to the owner.

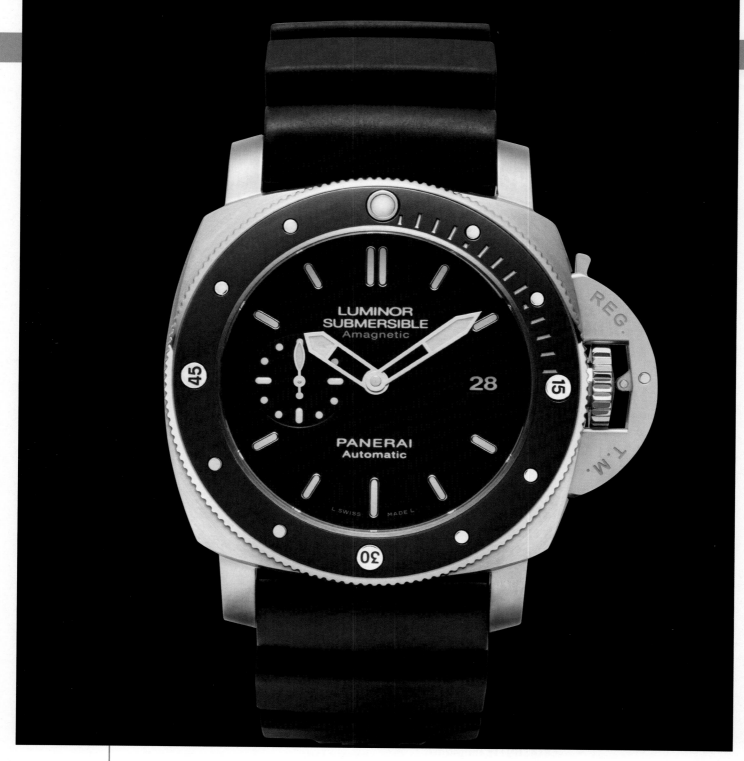

PANERAI

LUMINOR SUBMERSIBLE 1950 AMAGNETIC 3 DAYS AUTOMATIC TITANIO
REF. PAM00389

Endowed with an innovative case construction that provides a staggering degree of magnetic resistance—40,000 A/m—this 47mm brushed-titanium diver's watch accompanies its wearer deep beneath the surface with subsidiary indications of date at 3:00 and small seconds at 9:00. Armed with a unidirectional bezel, the timepiece achieves its extraordinary anti-magnetic distinction through the housing of the P.9000 self-winding caliber within an internal protective case crafted of soft iron of the highest purity. Immediately beneath the dial, the Faraday cage crucially diverts the flow of undesired magnetic fields, ensuring the optimal operation of the timepiece's mechanical core.

Brand Directory

A. LANGE & SÖHNE
15 Altenberger Strasse
01768 Glashütte
Germany
Tel: 49 35053 440
USA: 800 408 8147

AUDEMARS PIGUET
1348 Le Brassus
Switzerland
Tel: 41 21 845 14 00
USA: 212 758 8400

BELL & ROSS
8 Rue Copernic
75016 Paris
France
Tel: 33 1 73 73 93 00
USA: 888 307 7887

BLANCPAIN
6 Chemin de l'Etang
1094 Paudex
Switzerland
Tel: 41 21 796 36 36
USA: 877 520 1735

BREGUET
1344 L'Abbaye
Switzerland
Tel: 41 21 841 90 90
USA: 866 458 7488

CHANEL
25 Place du Marché St. Honoré
75001 Paris
France
Tel: 33 1 40 98 50 00
USA: 212 688 5055

CHAUMET
12 Place Vendôme
75001 Paris
France
Tel: 33 1 44 77 24 00

CHRISTOPHE CLARET
2 Route du Soleil d'Or
2400 Le Locle
Switzerland
Tel: 41 32 933 80 80
USA: 954 610 2234

DE BETHUNE
6 Granges-Jaccard
1454 La Chaux L'Auberson
Switzerland
Tel: 41 24 455 26 00

de GRISOGONO
176 bis Route de St. Julien
1228 Plan-les-Ouates
Switzerland
Tel: 41 22 817 81 00
USA: 212 439 4220

DIOR HORLOGERIE
44 Rue François 1er
75008 Paris
France
Tel: 33 1 40 73 59 84
USA: 212 931 2700

F.P. JOURNE
17 rue de l'Arquebuse
1204 Geneva
Switzerland
Tel: 41 22 322 09 09
USA: 305 572 9802

FRANCK MULLER
22 Route de Malagny
1294 Genthod
Switzerland
Tel: 41 22 959 88 88
France: 33 1 53 67 44 39
USA: 212 255 8499

FRÉDÉRIQUE CONSTANT SA
32 Chemin du Champ des Filles
1228 Plan-les-Ouates
Switzerland
Tel: 41 22 860 04 40
USA: 877 619 2824

GIRARD-PERREGAUX
1 Place Girardet
2301 La Chaux-de-Fonds
Switzerland
Tel: 41 32 911 33 33
France: 33 1 72 25 65 43
USA: 201 804 1904

GLASHÜTTE ORIGINAL
1 Altenberger Strasse
01768 Glashütte/Sachsen
Germany
Tel: 49 3 50 53 460
France: 33 1 53 81 22 68
USA: 201 271 1400

GUY ELLIA
16 Place Vendôme
75001 Paris
France
Tel: 33 1 53 30 25 25
USA: 212 888 0505

HUBLOT
33 Chemin de la Vuarpillière
1260 Nyon 2
Switzerland
Tel: 41 22 990 90 00
USA: 800 536 0636

IWC
15 Baumgartenstrasse
8201 Schaffhausen
Switzerland
Tel: 41 52 635 65 65
USA: 800 492 6755

JACOB & CO.
1 Chemin de Plein-Vent
1228 Arare
Switzerland
Tel: 41 22 310 69 62
USA: 212 719 5887

JAEGER-LeCOULTRE
8 Rue de la Golisse
1347 Le Sentier
Switzerland
Tel: 41 21 845 02 02
USA: 212 308 2525

LONGINES
Saint-Imier 2610
Switzerland
Tel: 41 32 942 54 25
USA: 201 271 1400

PANERAI
259 Viale Monza
20126 Milan
Italy
Tel: 39 02 36 31 38
USA: 212 888 8788

PARMIGIANI
11 Rue du Temple
2114 Fleurier
Switzerland
Tel: 41 32 862 66 30
USA: 305 260 7770

PATEK PHILIPPE
141 Chemin du Pont du Centenaire
1228 Plan-les-Ouates
Switzerland
Tel: 41 22 884 20 20
USA: 212 218 1240

PIAGET
37 Chemin du Champ-des-Filles
1228 Plan-les-Ouates
Switzerland
Tel: 41 22 884 48 44
USA: 800 628 4344

RICHARD MILLE
11 rue du Jura
2345 Les Breuleux Jura
Switzerland
Tel: 41 32 959 43 53
USA: 310 205 5555

ROGER DUBUIS
1217 Meyrin 2 Geneva
Switzerland
Tel: 41 22 783 28 28
USA: 888 738 2847

STÜHRLING ORIGINAL, LLC
449 20th Street
Brooklyn, NY 11215
USA: 718 840 5760

TAG HEUER
6A Louis-Joseph Chevrolet
2300 La Chaux-de-Fonds
Switzerland
Tel: 41 32 919 80 00
USA: 973 467 1890

VACHERON CONSTANTIN
10 Chemin du Tourbillon
1228 Plan-les-Ouates
Switzerland
Tel: 41 22 930 20 05
USA: 212 713 0707

VULCAIN SA
4 Chemin des Tourelles
CH-2400 Le Locle
Switzerland
Tel: 41 32 930 80 10

ZENITH
2400 Le Locle
Switzerland
Tel: 41 32 930 62 62
USA: 973 467 1890